The Medical Profession Is Dead and the Doctor Is "Critically ill!":

A Family Practice Memoire Exposing the Real Forces Behind Our Health Care System's Quality and Cost Issues

The Medical Profession Is Dead and the Doctor Is "Critically ill!":

A Family Practice Memoire Exposing the Real Forces Behind Our Health Care System's Quality and Cost Issues

Alan D. Cato MD, ABFP

ISBN 978-0-557-17882-7

Prologue

On my worst down days, I sometimes think the younger doctors have the upper hand when it comes to soldiering on in this business driven, paper-chase medical atmosphere of today. They should. After all, the young ones have never known any other medical environment for comparison. I have though, and, if that makes my sorrow at today's system an additional burden for me that my younger colleagues don't bear, so be it. Paraphrasing the old love saw, it is better to have experienced a grand profession and lost it than never to have known the grand profession at all!

Reader's from within the medical or political sectors, whose primary interest is only in the physician-author's opinions as to the current economic and quality issues plaguing our nation's health care delivery system, should fast forward and begin with **Chapter 4**. Readers should remember that the information, assessments and conclusions contained within this work are the author's personal ones, although based factually on observations and experiences through thirty-five years of practice as a family physician. The names of all persons and communities have been changed in the interest of anonymity and privacy. Throughout the work, the author has taken the name, Dr. Daily.

General readers looking for a fun medical read and a practical education concerning the pitfalls within our current health care system will want to begin with **Introduction,** for a colorful trip through a time when the medicine practiced in our health care system was effective, efficient, affordable and fun. Along the way, the reader will be made privy to an exploration of the many advantages of a past system, influenced by a proud *medical profession,* over the current one guided by big business principals and the quest for better bottom lines.

Introduction

The past few years have been the pits professionally, each successive year worse than its predecessor. My non medical friends tell me it's only burnout after thirty-five years as a physician. I know better though. Hell, compared with our predecessors, precious few of us today work long enough hours to qualify for burnout. When medicine still was a *profession,* our forefathers had deplorable hours and even worse physical working conditions, but they were happy with their work. They died young, but they were happy!

I'm sick at attempting to make years of medical education and the philosophy that each patient is different fit today's unyielding business mold of one policy fits all. It's unbelievable how far we've strayed from our mission of caring for the ill in an effective, efficient and economical manner. I am disgusted from watching on a daily basis: physicians practicing contrary to their scientific educations, doctors playing God (not by pulling plugs) but by plugging in devices for prolonging physiological existence but not meaningful life, once venerated hospitals now reduced to only recognizable corporation names, medical personnel more determined to follow a policy than to act spontaneously and save a life, patients demanding and receiving treatments adverse to their best interests, the RN title being educationally diluted in the case of the two-year degrees and educationally inflated in the case of primary care nurse practitioners, and persons with only a minute fraction of the education of that of an MD driving the cost of care up and safety down by essentially practicing medicine with minimal, to no supervision.

During the seventies I knew precisely what an RN was. If a person was addressed as nurse in the hospital setting or used the initials RN after his or her signature, I knew immediately that the individual was a valuable health care delivery partner whose education was sufficiently grounded in the medical sciences for communicating with physicians and for clinically assessing patients. Today, this is no longer the case. Sadly, due to the clamor from hospital management business corporations for more nurses and, with management corporations' focus on immediate quantity rather than quality, the concept of the two-year associate RN degree was able to be sold, changing the nursing profession's educational complexion entirely. Recently, I found myself called upon for settling an argument among three nurses by explaining why a tubal ligation did not eliminate menses. All three of these nurses were females! I also recently had a newly minted 2-yr RN ask, "Does an NPO (nothing by mouth) order include the patient's IV medications as well? I have driven to the hospital in the middle of the night to see a patient whom I was told, by a newly minted young nurse, was cyanotic (a blue hue to the skin due to insufficient oxygen levels in the blood), only to discover that the "cyanosis" was localized to an area of the patient's back and matched the blue color of the paper incontinence pad beneath him, rubbing off onto my hands at touching it!

All too frequently, I've watched with trepidation as many such hastily turned out nurses, uncomfortable in clinical roles because of their scant health sciences backgrounds, elect to get out of direct clinical responsibilities and use their new degrees for serving as quality assurance surveyors, Medicare surveyors, OIG surveyors or any one of the large number of other paper trail inspector positions, paradoxically giving those individuals, with the least clinical backgrounds, life and death input over the reputations of doctors, hospitals, and their clinically more astute nursing colleagues who chose to remain in direct clinical care. I have watched with even more concern as many of the BS degreed nurses in direct clinical care become disillusioned with the increasingly non clinical aspect of their jobs and with sharing the title of nurse with their less educated colleagues, consequently develop the illusion that they have more in common with physicians than with their two-year program nursing cohorts. Misled by such illusions and taking advantage of today's misinformed politicians, special interest groups and greedy medical management firms, I've watched with concern for the well being of patients as many of these disillusioned B.S. degreed nurses aspire to so-called "physician extenders" roles. The result being that they end up essentially practicing medicine without an MD degree and adding more hands into the reimbursement cookie jar in the process. Can this possibly be a good thing for patient safety, quality or cost control?

Today, mainstream media are increasingly focusing on news of the U. S. health care system, and, unfortunately, much of that news is bad. Outlandish health insurance premiums, millions of Americans without any health insurance, poorly trained and over-worked nurses throughout our nation's hospitals and thousands of hospital deaths annually due to mismanagement or medical errors are repeatedly being spotlighted by the media. Increasingly, many of these reports are far surpassing any reasonable definition of mistakes and smack instead of lack of integrity, education, management, supervision, pride and, most important of all, loss of <u>professionalism</u>.

Among the more bizarre *hospital mistakes* making the news recently are: a hospital in a big-name system accidentally using discarded elevator hydraulic fluid for sterilizing its surgical instruments, several clinics using a new endoscopy device and later having to contact hundreds of their patients to suggest they be tested for HIV, Hepatitis B and Hepatitis C due to their not having followed the manufacturer's recommended sterilization procedure for the new instrument and, one hospital, reporting operating on the wrong side of the brain three times in one year!

For the past few years, experiences like these and the daily occurrence of countless other insults to the once noble practice of medicine have convinced me that the medical profession is truly dead. I don't know its precise time of death, but I must have been at its bedside, since it was nothing like this when I first got on board in 1974. Looking back, though, there were clear signs and symptoms even then that the profession was already sub clinically ill. That's the purpose of this frustrated doc's rehash of his career and experiences, for placing some of these early signs and symptoms in print, so that anyone, with an interest in medical history and dissatisfied with today's medical care, can understand how the medical profession came to die, and, for the young doctors in the *medical business* today, providing them with historical proof that a noble medical profession really did once exist!

CHAPTER 1

I'm not sure of the precise point that I decided on becoming a physician. I do know that by my junior year in high school the decision had been made, although I kept it to myself, so unlikely my chances of success seemed. Usually, whenever my dream was publicized, it was through my father. Dad was a pharmacist, and I had plenty of opportunities growing up for noting the awe with which he regarded physicians. His drugstore, being in the same building as two physicians' offices, afforded me frequent contact with those two docs as they often took their brief breaks at the soda fountain, still a standard fixture in drugstores in the early fifties. This also gave me an excellent vantage point for observing the respect which other patrons of the drugstore reserved for the title of doctor. For a small boy fascinated with biology and whose Christmas list each year included a larger chemistry set than last year's, the chances were reasonably good that I would have gravitated toward a medical career even without my dad's subliminal and overt manipulations for assuring it.

Many years later, in the painful death-grip of medical school, I heard similar reasons for being there voiced by my classmates. Extremely significant in thinking back on it, during the entire four years of medical school I never once heard any of my colleagues mention lifestyle or money in referring to why they were there in pursuit of the MD degree! That was one of the truly amazing things about my medical class, the hidden underlying homogeneous nature of our motivations, aspirations, and goals. The first week there it seemed that we were eighty-eight persons of different religions, politics, social graces, and places of origin. We had Protestants, Catholics, Jews, agnostics and atheists. We had people with corvettes, old pickups, motorcycles, some who walked, and some who rode the bus. We had people dressed in the latest fashion, a few barely even dressed, and one or two who chose the then popular hippy fashion. We had quiet individuals, loud obnoxious individuals, serious individuals, nerd individuals, shy individuals, and individuals who loved teasing and preying on the personal traits of us all, but each of our reasons for being there was the same, a passion and fascination for the medical sciences and an admiration for physicians.

In the clique I eventually gravitated to, we had the epitomes of both these last two categories in the extraordinarily introverted and shy David Pixley and the extraordinarily extroverted agitator, Mike Huggins. David was the picture-poster shy country boy with his long sandy hair and a light complexion over which you could follow the leading edge of the shiny red wave of embarrassment whenever he was required to speak or, sometimes, even with just being spoken to. Mike was exactly the opposite. If he entered a room that was quiet or too serious, it wouldn't be for long. He was an outgoing teaser par excellence, and since, like the rest of our little circle, David was a favorite of his, he promptly dubbed him "Dave Dudley Pixley", almost immediately shortening it to simply "Dave Dudley," and it stuck for the entire four years. For those four nightmarish years of medical school, the pranks by Big Mike, as we called him, on our mutual buddy Dave Dudley, provided our primary source of relief and entertainment in an otherwise grueling academic hell.

Big Mike was innovative and dedicated to his teasing. I recall one time his arriving early at the study lounge our group frequented, and while the urinals in the men's room were still dry from overnight, painted on their gleaming white interior surfaces, "Private Piss Pot of Dave Dudley Pixley Esq." And over the toilet booth's door he had placed a very realistic plaque made of poster paper and done in old English style lettering, "Welcome one and all to the newly renovated Dave Dudley Pixley Esq. Memorial Crapper." As sophomoric as it was, it provided much needed and repeated relief from our stress for the next several months when, just by mentioning the need to use the Dave Dudley Pixley Esq. Memorial Crapper, Dave's face would glisten red and he would decompensate all over again.

Big Mike saved his best Dave Dudley prank for greater numbers to enjoy, not just our own little circle. Grand Medical Rounds were held monthly in the great old amphitheater of the University Hospital. These were always conducted with great decorum and included a nationally renowned visiting physician from one of the larger medical schools. Attendance at these venerable and august proceedings was required of all the house staff and all medical students in their junior and senior clerkships within the hospital. The format for these solemn events was a difficult or unique case presentation, followed by discussion among the faculty and a lecture from the visiting dignitary on the pertinent teaching points of the case. It was a stressful and nerve wracking time for all present, particularly for the students and residents currently assigned to the clinical department presenting the case, as they had to sit down front among the medical school faculty and were prime targets for questions from them and the visiting professor. These poor souls were in the greatest danger, but all in attendance had sweaty palms since occasionally the spotlight from down on the podium would select some poor devil seated higher within the amphitheater for a sudden question. Because of this we always entered that majestic old amphitheater with rapid pulses and nervous chatter and positioned our little group at the very top, in the shadows. There we worked at remaining as inconspicuous as possible. Even Big Mike's volume was toned down on these occasions. Arriving at one of these events even earlier than usual, we were ensconced in our preferred high altitude seating and nervously surveying the grand old theater, watching it slowly fill to capacity with white coated MD aspirants. Big Mike, in muted voice, turned up the pressure by claiming to have heard that this particular visiting professor was well known for singling out medical students rather than house staff for his questions. He kept reminding us to keep our voices low and, above all, do nothing that might call attention to ourselves. By the program's start he had us all whipped into near paranoia and thoroughly convinced that we were going to see one of our own get dismantled by the speaker's sharp mind and sharper tongue. It went well, however, and by midpoint in the proceedings no student or house staff blood had been let. The visiting dignitary physician was making a point regarding the number of diagnoses to be considered in the differential, and he paused a second for a faculty member to provide him with a piece of chalk. After a pregnant moment in which our embarrassed faculty member fumbled interminably in a box before finally providing him with one, he continued. Now as the visiting medical professor dignitary stepped up to the center most blackboard, the overhead lights automatically focused onto it, as all Grand Rounds were preserved on video for posterity. Looking up for beginning his list on the board, he noted that the large pull-down projector screen was in the lowered position and covering the center of his blackboard. Still speaking to the audience, he reached up and pulled down on the projector screen in order to get it to retract by its old spring-loaded mechanism and thus free up the blackboard behind. All adjacent blackboards were entirely free of any left over markings, but it was the center board beneath the spot lights that he was intent on using. Slowly the screen creaked upward after his releasing tug and, as it rose slowly upward, gleaming white chalked letters began gradually to come into view until an entire greeting was able to be seen by all. "Welcome to the 88th Dave Dudley Memorial Grand Round Lecture Series!"

As different as our personalities and backgrounds were, by the grueling four years' end it was clear to all of us that we were much more alike than different. Vocational choice, premedical academic competition, and the nightmarish grind of medical school had seen to that in the end. We were a group of twenty-two-year-olds who, since our fascination with the sciences in grade school and infatuation with physicians began, had one and only one focus in our lives. We continuously ruminated over it, hoped for it, planned for it, sacrificed social lives for it, prayed for it, and lived a decade of our lives in constant fear of not attaining it— the MD after our names.

By the beginning of college, all of us had investigated enough to be familiar with the harsh reality of our missions. We were all too aware that attainment of the MD degree required the longest and most punishing academic road of them all. Even had we not been aware of the difficulty of the academic road ahead and arrived naively at our freshmen year in college, our faculty science advisors would have brought us quickly to reality. I still remember my first meeting with my faculty science advisor as if it were yesterday.

"Well Mr. Dailey, what are you planning to major in here at good old M.U.?"

"Premed, Sir," I replied, glad for the chance of letting him know he was dealing with a serious student here.

"There's no such thing as a premed major, Mr. Dailey. I assume therefore that you mean you will be majoring in chemistry or biology with the hope of being accepted to medical school upon graduation from here," he droned in rote and uninterested fashion.

"Well, yes sir. I guess that's what I mean then," I mumbled.

"You and every other freshman in Chemistry 101 and Biology 101", he chuckled and then quickly added, "But after your freshman year you really should consider taking a few education courses as electives. That way you have something to fall back on. In other words, you can always use your degree for teaching if you don't get accepted to medical school."

"Yes sir, I'll think about it," I lied.

"I don't mean to discourage you," he continued, but you are talking about the most difficult graduate studies school of all for getting into. They are extremely selective. Even if you should graduate from here with the necessary grade point average for making application, they have literally hundreds of grade-point qualified applicants for each freshman seat in their programs. They can afford to be choosy." He paused a bit here and added in a softer tone, "I just wanted you to know what you are in for so that you don't let your guard down for even a second while you're with us. You can't afford a grade below a B, not even in physical education, and most courses had better be A s."

With that meeting began the first-half of a rigorous, punishing, and nonstop academic competition beyond anything that I could have ever imagined. The passion and focus of the competition were total and made us our own worst enemies where academic load was concerned. We must have been a science professor's dream, since we voraciously chewed up every assignment regardless of its demands, registering for increasingly complex graduate courses in biology and chemistry as the semesters crawled by. If a professor announced ten bonus points added to the final exam grade for reading an entire book and writing a ten page report on it, each of us read that book, never mind that the announcement was made only three days before the scheduled final! No matter how wiped out or how pressed for review time for other exams, passing up those bonus points simply was not an option, since ten extra points on a final exam was more than enough for knocking you down from an A or B to a death-dealing C on that dog-eat-dog grade curve. By the finish of these undergraduate years, the premed students began to develop a personality common to the group, characterized by extreme pride and self worth. This was undoubtedly due in large measure to our science professors' respectful attitudes toward us over the final year of the grind. They had had their way with us for four hellish years and no matter how demanding or senseless their challenges seemed, we met each and every one. We had to. Competition for those few medical school seats transformed us into the undergraduate school professors' academic slaves.

In the summer before my freshman year of medical school, I asked my own personal family doctor an incredibly stupid question.

"Is medical school really as bad as we hear?" He grinned as he was formulating his answer.

"Bob, did you really enjoy your last four years in college?"

"Are you kidding? What was to enjoy? I didn't do anything but study and worry."

"Before you went, could you ever have imagined the amount of effort and time that you would spend or the amount and complexity of the material you would cover?" Easy answer there,

"No way!" I admitted, falling further into his trap.

"Well then, remembering how you were surprised there, imagine now the worst ultimate medical school nightmare scenario you can." He paused here and then asked, "Have you got an idea now?"

"I guess so," I sighed.

"Well, now take that idea and times it by ten and you'll be somewhere in the ballpark as to what medical school is really like."

My family doctor was not being overly clever. If the pace of medical school was incredible, the volume of material was no less so. I still have a nine-hundred page pathology textbook that we covered in a little over one semester, and at least eighty percent of its pages are covered with faded yellow highlighter, indicating that I actually found time for reading it! Handwritten, nearly illegible notes from that one course filled five spiral-bound notebooks, not to mention a stack of preprinted lecture handouts filling six shoe boxes. The note taking came during class. The reading of the textbook and lecture handouts came in the evening. The studying and committing to memory of it all came after midnight, each and every night, with too few exceptions worth mentioning.

Today, I'm told that medical school classes hire class stenographers for relieving them of this onerous task of note taking, enabling them to concentrate on lectures and follow what's being presented to them at the time. This, along with advances in audio and video technology, must make today's medical students' lives almost bearable! Our generation did not have these luxuries, and this rapid-fire note taking led to our having notoriously bad handwriting by the completion of medical school, not the written-in-Latin myth so widely ascribed to by the public. You did not have to be a genius or an incredible theoretic thinker to get through medical school. You simply had to possess cognitive skills capable of assimilating, retaining, and regurgitating upon demand unreasonable volumes of material presented over unrealistically short periods of time. The volume and depth of material was due to most of the instructors being researchers who were presenting to us, in an hour or two's lecture time, data accumulated over ten to fifteen years of their lives, focused on one select and limited area of biological science.

So it went for the first two years of medical school, two additional years of even more basic medical sciences, designed for understanding in minutest detail the anatomy, physiology, and biochemistry that enable the human body to perform the extraordinary functions it does in health and for learning the disease states that can afflict it, including their abnormal physiological mechanisms. Only by possessing such a vast biological and chemical knowledge base is it possible to successfully diagnose and address an illness which you may be seeing for the first time, an illness that appears atypically from its textbook description, or an illness that appears atypical because of the presence of other concurrent illnesses in the same patient. Diseases occur in textbooks, but illnesses occur in persons. Pneumonia, for example, does not look the same in a thirty-year-old as in a ninety-year-old. The bacterium causing pneumonia in a healthy adult is not the one that therapy should be directed toward in the pneumonia of a cancer patient on chemotherapy, and the agent causing pneumonia in an AIDS patient requires therapy which will cover for even additional bacteria and other unusual organisms as well. The pain of appendicitis is not the same quality in an eighty-year-old as in a twenty-one-year old. The location of the pain of appendicitis in the eighth month of pregnancy is not the same location as in a six-week pregnancy. Disease and the practice of medicine are not static unchanging entities that lend themselves to rote memory for learning about. Both are in a constant state of flux and evolution. The best possible armament for coping with this fact is an inexhaustible medical health science background as taught in traditional medical schools. Most good physicians I've talked with regarding their medical education and its value to their every day practices, tell me they wish they could go back for even more of the basic science years, not for additional specialty residency years.

An education, consisting only of memorization of a few facts coupled with previously experienced clinical situations, is not sufficient learning method for producing quality physicians and certainly insufficient for turning out doctors capable of successfully addressing the unknown, unique, atypical, unexpected, or new. The most effective and economical weapon that the physician has at his disposal is the inexhaustible medical science knowledge base as traditionally and gruelingly taught in the medical schools of the U.S. When you pay a physician's office fee, you are paying for the distinct advantage that his/her education contributes towards your odds for an economical, efficient, and positive outcome to your illness. You are paying that physician for accessing his extensive education for concluding that your abdominal pain and vomiting during the local viral intestinal flu outbreak, when half the town has abdominal cramps and vomiting, has certain tell-tale signs that make it statistically likely to be appendicitis and worth the cost of some additional testing for being safe rather than sorry. Carrying out this screening and separating the likely appendicitis from the intestinal flu cases is deceiving to an observer of the process, since they only hear and see the doctor's conclusion and plan. This usually leads to their mistaken impression that the doctor is using a signs and symptom memory approach to the case. Memorization of clinical cases experienced in medical school simply is not sufficient for practicing quality medicine. The unusual will almost always be missed unless an extensive health science knowledge base is available for reasoning with. Take, for an example, an elderly patient with swelling of the feet and ankles. A clinician with limited knowledge of medicine gained only from observations while working with a seasoned physician would call on memories of past cases of edema of the feet and ankles for arriving at a diagnosis and treatment plan. Statistically, this means the bulk of those memories will be made up of the most dramatic cases, making congestive heart failure the diagnosis he most associates with swelling of the ankles. This approach to diagnosis leaves a host of other possible causes of ankle swelling not to even be considered in the differential diagnosis of a patient with swelling of the ankles. Liver disease, hypoalbuminemia (low serum levels of a protein), renal failure, pituitary disorders, congestive heart failure, and chronic venous insufficiency of the lower extremities all can lead to a patient having swelling of the feet and ankles. It depends entirely on the depth of medical science education of the person doing the diagnosing as to how many of these possible causes will even be considered in diagnosing a patient with swelled feet and ankles. As a junior medical student, the diagnosis of appendicitis is much easier made than it is at the end of specialty residency training. This is because in the junior year of medical school, the *only* thing the student knows of that causes right lower quadrant abdominal tenderness is appendicitis! At the completion of the residency years, the numerous other important conditions presenting as right lower quadrant abdominal tenderness have to be considered as well.

It's been my experience over the years that most nurses and some poorly trained physicians, relying only on memories of what they have seen done for a patient with lower extremity edema, think first of congestive heart failure as a diagnosis and thus order Lasix (a strong water pill) for initial treatment. A very few sometimes will consider bad veins in the lower extremities as the cause, but even then, they still think of a strong water pill as treatment in this instance as well. If, however, the swelling is not due to excessive water being retained as in heart or renal failure, a strong "water pill" will be of no benefit and potentially can be quite harmful, depending on the circumstances. Making a wrong diagnosis and subsequently choosing an ineffective or harmful therapy, is almost always due to the clinician not possessing sufficient physiological, biochemical and pharmacological knowledge base or not approaching the case with the physiological and biochemical mindset that is so necessary for successfully practicing medicine. This should give us pause for reflection today with "physician extenders", possessing only nursing education credentials, increasingly being permitted to serve as first-contact primary care providers.

Years after memories of specific cases demonstrated during medical school training have long sense faded, a physician can still call to mind all the possibilities for causing ankle and feet edema by accessing his basic science background for reminding him of each factor working physiologically to keep fluid within the blood vessels where it belongs and then considering each point within that system that failure could occur, particularly in the context of the specific patient before him. This method of approach to diagnosis is logical and simple but impossible to employ without the years of exhaustive medical sciences provided by the traditional medical school education process. This should be of concern to health care consumers today when special interests have lobbied successfully for legislation that permits nurses, having the scantest of medical science education relative to that of a physician, to diagnose, order expensive tests, and prescribe medications. I can't begin to tell you the number of times a year that, after evaluating a patient for swelling of the ankles and writing an order for supine bed rest intermittently during the day, elevation of the feet whenever seated, and TED stockings when weight bearing, I hear from nursing staff, "Doctor, do you think a fluid pill would help too?" Or, the number of times I'm called by nursing staff with the request, "Mrs. Smith is complaining of her feet swelling at the end of the day. Can we have an order for a diuretic (water pill) for a few days?" The frequency of such requests by many nurses today tell me that today's average nursing education does not even remotely resemble that of a physician's, and that frequently, nurses have not a clue as to how a physician's mind works or how integral the medical basic sciences are to the process's success. Yet, today the bulk of first time nursing employees being hired by hospitals are only 2-year degreed nurses, a group whose basic health science education has been shortened even more! Even more frightening is that, thanks to special interest groups within business and nursing, B.S. degreed nurses, thru undertaking several months more additional education at the hands of nursing instructors (today some programs are even providing a lot of the education material thru on-line courses), are legally being permitted to essentially practice as physicians with minimal or in many instances, no supervision.

As first-year medical students, my class was also ignorant of the contribution of basic medical sciences to our future abilities as physicians. After four years of nothing else in undergraduate school, we were sick of the basic sciences and anxious to begin working with patients. Our wishes, of course, were of no concern to anyone but ourselves during the four grueling years of medical school, and our instructors in basic sciences in medical school proceeded to show us that in our undergraduate years, "We hadn't seen anything yet".

A particular memory that points out the intensity, pace, and volume of those two basic science years, was the afternoon before our first test in medical school. We had all been on pins and needles for the preceding two weeks in anticipation and dread of our first test as medical students. All had already begun systematic reviews of notes and materials several weeks before. Because of the volume and depth of the material covered, review and studying for an approaching test was always begun at least a month in advance. Now, on the afternoon eve of the test we were totally wasted, since most had made our already late night review hours even later as test time drew nearer. Usual strategy called for no further study the night before an exam and getting as much sleep as possible, beginning immediately after our last lecture of the afternoon. The last lecture of the day before our first test as medical students was in anatomy, and we were looking forward to entering a new system of the body, the central nervous system. Starved for sleep as always and just released from an hour in a darkened room with one teaching slide after another presented and commented on in the nasal monotone voice of the professor, I hoped I was having auditory hallucinations when finally switching off the projector I heard him say,

"You might want to consider cutting your break short and get on to the lab for an early start, since today's lecture and lab material will be included on your test tomorrow." Stunned, we bypassed the break entirely and hurried directly to the lab to see for ourselves. On the tables there were stone crocks of formalin, each containing an entire human brain. Streaming in tangles over the edges of the crocks were threads. Each thread was stapled to an index card dangling near the crock's base. On each card was the anatomical name for the area of the brain being pierced by a pin attached at the thread's other end. Adjacent to each crock was a shallow tray of formalin with slices of brain from front to back, lying in sliced bread fashion in order for making visible the anatomically identifiable tracts within the brain. Similarly, surrounding each tray was a tangle of threads and index cards, each leading to an individual pin stuck somewhere into one of the slices. Someone did a quick count and came up with one hundred forty additional potential questions for the next morning's test! Normal individuals would have protested and refused en block this unreasonable demand and gone on home for a good night's sleep in preparation for our first big test. Medical students back then were not normal individuals. They would never have gotten to be medical students if they were. So, about nine o'clock that night benches began scooting and crock lids clanged closed, and midst much grumbling and griping, we all left the building to sleep for what remained of the night.

Medical school's clinical years were no less stressful and demanding, but in a different manner. Fear, fatigue, and embarrassment were the toll takers of the clinical years. We feared that a patient would suffer harm due to our inexperience. We feared being grilled on rounds by the attending physician. We feared we would become bumbling klutzes if called upon to perform some procedural skill on rounds. Most of all, we feared appearing inadequate in front of our fellow classmates. We feared the professional literature we referenced during case discussions would be pointed out by a competitive classmate as outdated. Academic competition had not ended at the medical school doors. It was alive and well in the hospitals, too. In fact, it was even greater, although in different form and with grades no longer the entire goal. The additional new goal was *professional respect* from peers and faculty. The primary weapons of this intellectual battle became professional journals instead of textbooks, and the primary challenge became finding time for ferreting out the most current of these while keeping up with your hospital duties as well. This all-out quest for the admiration of one's clinical acumen by one's peers was the defining characteristic of the clinical years and, *more than anything else*, formed the basis of the integrity and character of the *medical profession* until its untimely death at the hands of bureaucratic and business interests. Finished with the interminable schooling and out in practice, doctors of past times continued to seek the respect of their colleagues. "A thinker", "a real doc", "a doctor's doc" was a term in the professional vernacular that each wanted to think of his colleagues using when speaking of him/her to other physicians.

Regrettably for physicians and patients alike, today's medical system influenced by big business, terrorized by fear of litigation, and focused on the paper trail has had a cookbook approach to the practice of medicine foisted upon it by the trend toward clinical guidelines and clinical policies being developed by hospital administrations, risk managers, and regulatory agencies. In the process, the individual doctor's personal motivation and responsibility for striving for excellence above that of his peers has been amputated. Indeed, the current system seems geared toward leveling the quality of all physicians to the same standard, and that standard is average. This is being done for administrative convenience and through the fallacious thinking that medicine practiced by protocol is effective, error proof, and malpractice proof. Policies and guidelines lend themselves well to the paper trail worshiped by hospital management for its ability to make the government regulatory agencies' jobs easier, but these same policies lend themselves very poorly to individualized patient care.

For patients, this trend has resulted in the near extinction of one of their most convenient, cost-efficient, and effective means of health care delivery: the independent, highly educated, professionally motivated, tell-it-like-it-is, personal physician acting on clinical judgement and not playing the defensive medicine game with needless tests and documentation.

There was a time when good clinical judgement was the standard of care and failure to exercise it was the most common cause for the rare litigation event. If a physician had a record indicating that the history and physical exam suggested no reason for ordering a CT scan of the head and indicated the patient was suffering a muscle contraction tension headache, there was no danger of litigation if that patient years later under different and unrelated circumstances, died of a ruptured aneurysm of the brain. Today, some attorney might successfully convince the family of such a patient that if a CT of the head with contrast had been ordered years earlier when the patient was being seen for his tension headache, the aneurysm might have been *coincidentally* found, and the patient might have lived. This might well be true, but what is not true is that the doctor was guilty of malpractice. In actual fact, he may have practiced excellent medicine, showing by history and physical exam that the headache the patient was complaining of on the day he was seen by the doctor was indeed, a tension headache. In fact, the doctor was practicing responsible medicine and responsible health dollar economics by not ordering a CT of the head under such clinical circumstances. Today, increasing numbers of irresponsible physicians are concluding the patients they are examining have muscle contraction headaches but going ahead and ordering a CT of the head for the purpose of protecting themselves from possible unjustified lawsuits i.e., practicing defensive medicine.

A significant number of today's lawsuits are filed on the basis of bad outcomes. There have always been and always will be bad outcomes, but bad outcomes are not synonymous with malpractice. Unfortunately, attorneys have a proven success record of convincing juries otherwise, and it would be more accurate for the exorbitantly expensive insurance policies which physicians are forced to carry today to be called bad-outcome insurance, rather than malpractice insurance. Whatever the insurance is called, it is outrageously expensive and as overhead, is passed right along to each and every patient in the form of higher doctor fees and higher insurance premiums. If the public wishes to preserve its right for suing for bad outcomes in addition to malpractice, and if attorneys wish to continue reaping unreasonably inflated settlements for settling bad-outcome suits, it would be fare and beneficial to the public, as well as to physicians, that one-half of all awards in such cases go to a common pool for paying insurance premiums for physicians. Monitored through accounting processes this saving could be made to diminish fees and insurance premiums for patients.

The fatigue factor of the clinical years of medical school speaks for itself. Twenty-four to thirty-six hours without sleep were not at all unusual. Living out of a shaving kit and minimal bathroom facilities added considerably to the sense of fatigue. Embarrassment was the third characteristic of the clinical years. Medical students were the lowest in the pecking order. Even the experienced nurses had it over us. We didn't help ourselves much since we asked enough stupid questions, broke enough equipment, and contaminated enough sterile fields for satisfying any reasonably sadistic person's need for ridiculing, but that wasn't enough for them. We often found ourselves the victims of deliberate snubs, pranks, hurled insults, and even hurled instruments on the occasions that we really did show our ignorance in a manner that caused some higher-up an inconvenience or threatened a patient's well being. All in all, the clinical years were every bit as stressful as the basic science years and a whole lot more demeaning.

There were occasional enjoyable minutes during the clinical years though, and they usually involved a particularly memorable physician professor and always, of course, some one other than one's self as the medical student fodder. One of these interludes for me involved the chief of pediatrics, Dr. R. Blumbauer. Dr. Blumbauer was known for two things, his love of his own high I.Q. and his love of babies. His personal theory on training the perfect pediatrician was to begin with a high I.Q. medical student (never higher than his own, of course), have him or her memorize every major textbook of pediatrics and spend twenty hours a day in the immediate presence of a baby for the entire duration of their pediatric rotation. He steadfastly adhered to the belief that anyone wishing to learn about babies needed to observe their every function, under every circumstance imaginable, and in as much detail as possible. Each and every medical student at our school knew this about Dr. Blumbauer. We also knew his proclivity for precise definitions and his pet peeve of using the term "diarrhea" for describing any loose stool. Forewarned being forearmed, a student responsible for doing the admitting history and exam on a dehydrated three month old infant with diarrhea thought himself ready for Dr. Blumbauer on our rounds one morning. The student began with the baby's past history and a very detailed parental account of the infant's current illness and was doing admirably well.

"On the third day of the baby's temperature, he began having diarrhea stools and ..."

"Whoa, son," interrupted Dr. Blumbauer exactly as anticipated by us all. "Now just precisely what makes the stools of this little baby qualify as diarrhea?" Unruffled, the student reeled off the weights of soiled diapers, number of stools per twenty-four hours, water content, and so on, until he had reiterated precisely the standards as taught by Dr. Blumbauer, a stool needed to meet in order to be granted the title "diarrhea". Dr. Blumbauer continued looking at the student in silence for a second or two and then pushed some more.

"Did you take some of the stool and rub it between your thumb and index finger to determine its consistency?"

"Yes sir," the quick reply.

"What did it feel like, then?"

"It was soft and greasy feeling."

"Greasy, huh, Well then did you put some in a cup of water to see if it floated?

Again came the confident reply, "Yes, sir."

"What did it do?"

"It floated, sir."

"Well then, did you spread some of the stool out on some white paper to see if any subtle colors were evident in its liquid content?"

"Yes, sir, I did that too."

"Very good, Mr. Gardner, you must be smarter than you look. Were there any colors, then?"

"Not that I could appreciate, sir," he responded, now with an increasing confidence.

"Mr. Gardner, did you smell of that infant's diarrhea stool?"

"Yes, sir, I did."

"And what did it smell like?"

"It smelled like normal breast-fed baby stool to me."

"Ok, Mr. Gardner, let me ask you this last thing then. Did you taste the stool to see if it was salty or sweet?"

A second or two of silence ensued while the sniggering died away, and then came a disgusted and emphatic,

"No, sir, that I did not do!"

Turning and shuffling on toward the next patient's crib, head rotating negatively from side to side, indicative of his profound disappointment, Dr. Blumbauer called back over his shoulder,

"Well, son, it's your tuition money, it's your tuition money." It was part of the training and part of the tradition. The medical student could never be permitted to appear perfect on questioning by faculty on rounds or to have the last word.

Chapter 2

Now that I was on the final lap before entering the rest of my life as "Dr. Dailey," I was intently interested in picking out the ideal Kentucky small town for embarking on my adventure and dream of the first third of my life. I had always pictured my professional role as best being carried out as a small town doc in Kentucky, where I could continue to follow the basketball teams from Western Kentucky University, University of Louisville, and University of Kentucky, in any spare time I might be left with. Leaving no stone unturned in my search for my dream spot, I used every free opportunity during my internship in Evansville, Indiana, for crossing the Ohio River and investigating one potential community or another. As the man said though, "Life is what happens around you while you're making plans for it," and when I finally drove across the river in a packed moving van to my new profession and new hometown, it was not the Ohio River I crossed but the Wabash.

Physician life for me began in a southeastern Illinois town called Levelton. It was a community of some seven thousand people, located just across the Wabash from Indiana. Levelton had a picture-perfect main shopping street, a good economy, and high average education level. The senior medical staff there was indeed senior; two were in their mid-fifties and three others nearer seventy, and all of them had been practicing medicine in that town since they left their respective medical schools. For me, at age twenty-seven, it was truly baptism under fire. In town less than a week, I attended my first medical staff meeting at the hospital. The age of the building and the age of the staff around the table made the small room and its occupants seem more like a meeting of the officers of the local Independent Order of Odd Fellows (IOOF) than of an active medical staff. I was introduced by Dr. George, the chief of staff and one of two fifty-year-olds in the room. His age-peer appeared to be napping with his head cradled in his arms that were folded on the edge of our table. The meeting got quickly down to new business on the agenda since there wasn't any old business, and at this point two gentlemen who were seated side by side and had been quiet to this juncture took over the floor. Both were in their late thirties, and I learned they were physicians from a neighboring small community who occasionally used our hospital for their patients. As they proceeded I thought gratefully, *here at least was hope for some academics and continuing education among this aged medical staff*, since they were relating what they had learned at a recent review course they had attended. Specifically, they had learned of the practical uses of serum osmolalities when treating patients with fluid and electrolyte disturbances and were now hoping to get the medical staff to vote in favor of requesting the hospital board's approval for purchasing the instrument necessary for performing this function. They both were knowledgeable of their topic, and it was evident to me that these two docs were up to date and liked to remain at the top of their games. They adeptly stated their case as to why we needed this additional technology and sat back down to listen to their colleagues' reactions. They didn't have a long wait. One of the seventy-year-olds, a balding and rotund gentleman with squinting eyes and a dour expression, immediately mumbled that he didn't see any use in this as,

"We've taken care of patients with electrolyte imbalances for twenty years without it. Why do we need it now?" This view was promptly echoed from yet another member of the seventh-decade trio. This doctor was disheveled and unkempt, wearing fashion abusive dress pants that had a tell-tale slick shine in their seat and a white dress shirt on at least its third consecutive day of duty. His comment was terse and to the point.

"I wouldn't have much use for it," he muttered. Sensing the negativity and trouble, one of the petitioning young docs quickly reviewed the uses that could be made of the instrument and ended his plea with a pointed observation.

"I believe it's you guys who are always complaining that your patients want to be referred to the medical center in Dalsville when they need hospitalization. Patients react this way because they view our hospital as outdated and as not having proper tools to work with," he continued in an emotionally crescendoing voice. "And you know what?" he challenged. "They are absolutely right. There is nothing sadder than a physician who is trying to provide quality and up-to-date care for his patients but looks like a fool because he lacks the proper tool," he concluded. A few seconds of silence hung in the air, reflecting the surprise at the emotion expressed in his appeal. Suddenly then, from beneath the cradled face of Dr. Saxon, heretofore thought by myself to be sound asleep, emanated a shrill but slurred cackle followed by,

"Yes, by God, there is something sadder than a doctor looking like a fool because of not having the proper tool," he slurred, with spittle hanging in the corner of his mouth. "And that's a fool with his tool that doesn't know what to do with it," he blasted out with a slap of his hand on the tabletop. With his opinion now expressed, he resumed his previous position in total silence. His input broke the tension, and those of us under seventy enjoyed a good laugh. Even the seventy year olds smiled or chuckled. We did not get the equipment.

Over the next few years, I would learn much from and about this group of docs but nothing that ever made me doubt their dedication to their patients and profession or their gluttony for punishing working hours. Later that evening, I learned that Dr. Saxon was not sleeping at all but under the influence; an impaired doctor in the vernacular of the millennium. Dr. George explained to me that Dr. Saxon had struggled with his problem for years and that, "It just happens every now and then and we keep a really close eye out for it, and if it's happening we confront him with it and then take him home. In a day or two he's fine again. He really is a good doctor otherwise," he concluded.

I had several opportunities for validating that description of Dr. Saxon, and I can only say that when sober he was not just a good doctor. He was downright sharp! His patients dearly loved him. As the new doctor in town, whenever Dr. Saxon was out of commission with his "problem," I received the brunt of his patients in my office. I also was called to the ER for his patients during his absences more than the other doctors were. It chagrined me and remains a mystery to this day how often his patients would chide me for "never being around when I need you." Despite a recorder on my home phone, a beeper on my person, and a physician taking my calls on my days off, I was never able to satisfy Dr. Saxon's patients on the availability issue. This happened in spite of the fact that, as far as any of us knew, whether out of pocket or working, Dr. Saxon never left assigned coverage or a recording directing his patients on his home phone. Sadly, after my third year there, I had to live with Dr. Saxon's critical patients on a permanent basis, as one Fourth of July afternoon he shot himself in the head. He was in his mid-fifties and had voluntarily served as a physician in Vietnam.

These were hard working times in both the office and the hospital. There were three of us who shared office space and shared on-call duty. One of them, Dr. George, averaged seventy office patients a day in addition to doing two to three major surgeries in the hospital before coming to the office. The other senior associate in my office, Dr. Moore, assisted Dr. George with his morning surgery cases at the hospital and averaged fifty-five patients per day at the office. As the newcomer to our office, my own average was forty-five patients per day after only about six months there. I felt quite well worked, especially after taking my turn at weekend call. Apparently though, not everyone thought me the dynamo that I did. For a number of reasons, I refused to make house calls, disliked assisting in surgery cases, and did not do OB. I was vaguely aware that my own office associates found this unusual, if not downright amusing. After about nine months in practice, word came that a well-known octogenarian physician from a nearby town had died. It was announced that our offices would be closed in order to attend the funeral of this venerable colleague. It was also strongly suggested that I should consider attending the funeral, as nearly every physician from adjacent counties would be there and it would be the politically correct thing to do. Hating social functions in general and funerals in particular, I dreaded the thought of walking into that church, especially since I was about five minutes late and had missed the en masse seating in the first three rows of pews of all the physicians. Entering the sanctuary, I hesitated briefly to scan the crowd for seating space and glimpsed Dr. Moore rising slightly from his pew to catch my attention and motioning for me to join him. The sanctuary was entirely quiet at this moment, and as I walked silently toward the front to join him, I noticed that he was seated with and caretaking for an eighty-five-year-old retired surgeon whom he had checked out of the nursing home in order that he might make an appearance on this solemn occasion. I had never met him but had heard many stories of his legendary work capacity in his heyday. While I was still several rows away from them, I heard Dr. Moore explaining to this patriarch of the medical community who I was.

"That's Dr. Dailey, the new young doctor that's come in with us," I heard him say in a deliberately amplified voice to overcome the old surgeon's hearing deficit.

"Oh yeah, he's that new young fellow who doesn't deliver babies and doesn't want to work too hard, isn't he?" The old doc bellowed out.

With the three of us sharing weekend call, it was nice to have two consecutive weekends off; but beware when that third weekend rolled around. It was a nightmare! We had a two-story house, and my grown children still recall that on Friday evenings at whatever time I made it home, I sat in a recliner in the TV room downstairs with a telephone in my lap. Other than bathroom trips, emergency room trips, and rounds at the hospital on Saturdays and Sundays, that's where I remained till leaving for the office on Monday morning. It would not be exaggerating to estimate that from Friday evening until Monday morning, the phone would ring on average once every half-hour. The rest of the family would retire to the upstairs, turn off the phones there, and watch TV in their rooms.

The workloads that physicians had then would not be sustainable in today's medical arena. The now-popular trend of litigation and chasing paper trails and their accompanying paranoia and the loss of a trusting relationship between doctors and nurses would not permit it. Back then, though, it was possible, and physicians of different ambitions and motivations than most physicians today, along with better clinically trained nurses than today's nurses, worked hard together for their patients and genuinely had fun in the process.

Mrs. Hastings was a labor-intensive patient for my nurses and me. She was well known to all the other physicians in the community, and I naturally inherited her because I was the new doc in town. A severely and chronically mentally ill patient, she had been in and out of the state psychiatric hospital many times. She was only in her late forties, but she appeared considerably older. After all these years, her two defining characteristics branded into my memory were her grating, monotone voice and the repetitive message it delivered. Although Mrs. Hastings was not physically infirm, I can still see my nurse at the lady's side, one hand cradling her elbow and forearm as she inched her down the hall toward the exam room. Each encounter with her was identical. At the exact instant our respective positions permitted, her eyes would snap into a riveted stare onto my own. There they would remain, unwavering and unblinking, until I was compelled to break free of their grasp by glancing at her chart or my nurse. Whenever my gaze strayed back toward the vicinity of her face, however, those eyes immediately clamped like a spring-loaded trap onto mine, and the process repeated itself. Her eerily grating voice was in perfect synchrony with the snapping fixation of her gaze.

Even today, whenever I think of Mrs. Hastings, I can still hear that voice as if she were standing before me. It was a low, droning monotone, with the emphasis definitely on monotone. Inevitably and incessantly that voice communicated the same message and, unless interrupted by a question from me, was repeated over and over like a mantra. With her eyes glued to mine and her face absolutely devoid of expression she would drone,

"Oh, Dr. Dailey, I just want to die. I just want to die. Nobody cares about me, not even my family. Oh, Dr. Dailey, I just want to die." Only when I addressed her would the droning voice temporarily cease, but the precise instant that my own voice ceased, that vexing mantra began again. By repeating a question twice, it was occasionally possible to extract a yes or no reply from her, and that too was always in the same monotonic, grating fashion and without even a flicker of facial expression. Once she was ensconced in the exam room chair, aside from physical prompting and assistance, it was impossible to get this little lady to end a visit. What little I learned about her illness and past history I garnered through telephone conversations with her one family member still willing to discuss her, and I also gleaned some information through the state's psychiatric hospital. Neither of the two sources were much help, and the state psychiatric hospital staff assured me that on her last visit to them no indications were found for justifying her admission there. They also informed me that she would not take any kind of medicine, a fact we were all too aware of already. And so the months passed, and we developed a routine in the office for accommodating Mrs. Hasting's compulsive visits. Having three exam rooms, we simply placed her in one and I, after failing miserably to convince her to try medication or to see a psychiatrist, would leave her in that room while continuing to see patients out of the other two. Utilizing this method, we were able to accommodate her reluctance to leave the office and thus were able to save her and ourselves the embarrassment of an assisted exit through a full waiting room. It was sad and frustrating, but we had exhausted all outside resources, including local social services. When finally she did leave the exam room, accompanied by my nurse, I could hear that dreadful, droning death wish growing gradually more distant as the two of them shuffled slowly down the hall to the waiting room door. The distance was only twenty feet, but it seemed to me like an hour's journey was required each time before the closed waiting room door rid my brain of that voice and its message,

"Oh, I just want to die. I just want to die. Nobody cares about me, not even my family. I just want to die".

Long after she had gone, it seemed as if she had left her mantra behind, as we all kept hearing it in our minds. Everyone in the office had his or her imitation of Mrs. Hasting's chant, but none achieved the chilling and grating quality of the original. Among ourselves, we had often expressed our personal feelings that this poor soul might indeed be better off in death, given the degree of torment she labored under in life.

One Saturday afternoon I was passing through our hospital emergency room on my way out of the building when I heard the radio crackle out that an ambulance was en route to the ER with a gunshot wound to the head victim. Immediately the ER nurse blocked my path and asked if I would hang around to handle it. Within just a very few minutes we heard the sound of the ambulance in the drive and we met them at the double doors where two ambulance attendants were in the process of guiding the gurney through. The body on it appeared to be totally covered with a sheet and I guessed, "Didn't make it", as I helped pull the gurney on through the doors.

"No, Doc, she's still going. The sheet just got dragged over her head during the transfer, but I don't see how, I mean it was point blank to the center of the forehead!" At this juncture I had pulled the sheet down and was leaning over the head looking down at a stack of blood soaked gauzes covering the center of the forehead and extending down to partially cover the eyes as well. Gently lifting them away in order to examine the wound, both eye lids snapped open exactly as they do in the horror movies when the heroine is leaning over the vampire in his coffin, and simultaneously I heard,

"Oh, Dr. Dailey, I just want to die. I just want to die. Nobody cares about me, not even my family, I just want to die."

On that afternoon Mrs. Hastings decided to add action to her monotonous words and had shot herself in the middle of her forehead with a .22 caliber pistol. Fickle fate, an aberration of ballistic physics, and the small caliber had combined to thwart her efforts. Skull x-rays showed the small slug to have divided into three smaller fragments which then followed the surface of the skull in three separate directions, just beneath the scalp but never even entering the outermost layer of the skull bone! Although not in the manner she had intended, Mrs. Hastings had, by her action, bought a ticket toward her eventual relief via forced medication at the state psychiatric hospital. Thanks to the paper trail chase, litigation fears, and changed nursing attitudes the only physician you could get to voluntarily wait around the doctor-unfriendly hospital of today for the arrival of such a patient would be the paid and obligated ER doctor.

Dr. Kelly was the eldest of the seventy-year-old trio, probably nearer to eighty. He had a very large practice most of whom were senior citizens themselves. He had long since attempted cutting back on his workload by limiting his hospital admissions to those with more routine management requirements, and he no longer took morning office hour appointments, beginning instead about 2:00 p.m. He was a likable, slow moving, set in his ways, frumpily dressing man of simple needs and simple solutions. He socialized not at all, preferring when not in his office or the hospital to be in his own home. His typical workday did not begin until about 11:00 a.m. and then not with hospital rounds but with a hospital cafeteria stop for what could best be described as a country-brunch. He favored desserts, starches and anything with gravy on it, the latter fact which had taken its toll on his waistline and the colors of more than a few of his neckties over the years. From the hospital cafeteria it was up to the wards for a few minutes of chit chat with whatever nursing staff was around and finally a slow shuffle down the corridor to see his patients. I suspected he might not be the most up-to-date doc in the world, but I never heard anything from hospital staff or his patients that was problematic. I liked Dr. Kelly. What was not to like about this obese, modern day version of "Ole Doc" of *Gun Smoke* days. In fact, I often went out of my way to engage him in conversation, as I had been told that he looked at newcomers in the medical community as threats to his practice size and I wanted him to know that I was an alright guy. Probably because of such efforts, one Sunday morning I was surprised at getting a call from him asking if I would see any of his patients that might show up in the ER until he returned that evening. Pleased that he would even ask, I assured him that I would.

Less than an hour later I was called to the ER to see one of his patients who had been in a car accident. Our excellent nurse staffing the ER in lieu of a full time physician met me at the door filling me in.

"I can't really find anything on him except some major bruising, but he absolutely cannot stop coughing." From the nearest exam room long and violent paroxysms of coughing emanated, nearly nonstop. No more than fifteen seconds of quiet intervened between them and both the inspiratory and expiratory components were drawn out, much as we've all experienced when a small amount of liquid surprises us and ends up in our wind pipe instead of the intended esophagus. Despite the incessant coughing, upon entering the room a brief inspection of the guy reassured me. Between his paroxysms of coughing, this middle-aged man appeared in no other distress and was perfectly alert, complaining only of some discomfort over his sternum, not surprising since it was markedly bruised from impact with the steering wheel. My own ER training days not that distant in history, I was anxious to demonstrate and teach the nurse what I knew. In a nutshell I was a fresh young doc full of him self and wanted to impress the nurse with my worth to the medical community. In taking the patient's history I pointed out all significant worrisome clues including speed of the vehicle and steering wheel imprint positioned directly over the patient's sternum. During my exam, I correlated the steering wheel's imprint on the sternum to possibly injured anatomical structures beneath it, pointing out the very real potential for underlying injuries to the lungs, trachea, and heart.

Gratified that I had properly evaluated Dr. Kelly's patient and demonstrated the new doc's worth for the nurse, I looked at the clock to find that over two hours had passed, and it was now well into the afternoon. Still smitten with my teaching points for the nurse, it suddenly hit me how I could subtly share them with Dr. Kelly and thereby help establish my reputation among my peers. To carry this out I would have to do something very unusual for doctors in those days, dictate the history and physical immediately after performing it, so it would be available in black and white for him to see the depth of evaluation I had given his patient. Decision made, I retired to the medical records dictation cubicle which as it turned out, was immediately adjacent to the room that Dr. Kelly's patient had been admitted to. Throughout my dictation I continually heard the prolonged coughing paroxysms issuing from next door, and, had I only that sound to go by, I would have worried for the patient in that room. Just finished listening to a review of my dictation, I was still sitting quietly in the cubicle, imagining how thorough and impressive my work-up would appear when typed and placed in the patient's chart for Dr. Kelly to see and hopefully, mention to some of our colleagues or at least some of the nursing staff. Before I could get out of the dictation cubicle unseen I heard the approaching voices of Dr. Kelly and the ER nurse. Apparently he was back in town and had entered the hospital through the ER entrance for the purpose of seeing if anything had happened in his absence. The nurse was doing most of the talking and Dr. Kelly's usual pace gave me ample time for eavesdropping. She reviewed all the potential conditions that I had been concerned with, even informing him that I had ordered follow-up EKG's and cardiac enzymes for later in the evening. She continued her account of all that we had worried with uninterrupted, since old Dr. Kelly didn't comment a single time. Coming within hearing distance of the violent coughing ensuing from the room I heard the nurse's voice.

"See, that's him you hear now. Have you ever heard anyone cough like that so long at a time?" she asked the old doc. Still there was no comment from Dr. Kelly, so she added as they reached the patient's door, "That's why Dr. Dailey did several x-ray views, looking for free air from a tracheal injury."

There then ensued some unintelligible muttering in the room and after only a few feet more of shuffling across the room I heard Dr. Kelly calling out a greeting to the man. During the next cough free interval, I heard Dr. Kelly mutter another short phrase to the patient but could not make out its content. With that, I heard the door open and close and the two of them continuing on down the hall towards the nursing station without further conversation. Then after several seconds of complete silence except for the sound of their footsteps, I heard Dr. Kelly's voice again.

"Give him a little cough syrup every four hours," he instructed the nurse.

Not many days after Dr. Kelly's cough syrup patient, passing by the hospital administrator's office, I noticed my two associates, Dr. Zelle, and the hospital administrator huddled like hallway gossips just inside the door. Catching my glance as I passed, my partners motioned me to join them. Their lowered voices and the presence of Dr. Zelle clued me immediately that something was afoot, as Dr. Zelle was the dower faced 70-year old who had helped Dr. Kelly axe the request for a more modern "tool" at my first medical staff meeting. Like Dr. Kelly and Dr. Saxon, he was in a completely solo practice and looked out for only himself in all internal matters in the hospital.

"Come on in here, Dr. Dailey. We've got some problems you might be able to help us with. Anyway, you need to be aware of them," our obsequious administrator invited. They filled me in and the gist of it was that recently nurses and others had been voicing more concern over Dr. Kelly's orders. In short, they were questioning whether his age was finally affecting his patient care. When asked for my opinion, I had to admit that in the last month I had heard such talk among the nurses and also had one personal experience that seemed to substantiate the talk. Since my associate Dr. George was chief of staff, I had already informed him of the incident at the time it happened, and he urged me now to share it.

"Dr. Dailey, you seem to be the only one with anything other than hearsay. Why don't you share with all of us what happened with you and the pediatrician over at the center last week." Uncomfortably aware of the advanced ages of the other docs in the room, both within five years of Dr. Kelly's age, I related the event to all present, attempting to emphasize my own involuntary and passive role in the process. I told them that last week I had received a phone call from a pediatrician in the medical center in Dalsville. I explained to them that I had interned at the medical center in Dalsville and, while there, had established myself as having a special interest in pediatrics, pointing out that this was likely the reason behind the pediatrician there giving me the call. Since completing my training there, my interest in pediatrics had continued, and now, being in charge of our county health department's well-baby immunization clinic, whenever the need arose, I referred difficult and interesting cases to Dr, Quail, a pediatrician on staff at the center in Dalsville. I went on with my story, explaining that on the day my colleagues were urging me to tell them about, I had taken a call coming into our offices and immediately recognized Dr. Quail's voice greeting me. After the usual light banter, his voice took on a more serious tone as he said,

"Listen, I need a favor from you if you are willing to help. I really don't know if you can do it or not, because it doesn't involve your patient." He quickly proceeded to relate to me that while seeing a baby from our town in his clinic, the baby's mother had expressed to him her concern that the baby's cousin was lying over in Levelton's hospital "really sick, maybe even dying and just getting shots." He apologized for getting me involved, but the mother had seemed reliable and quite upset, and he had promised her to look into the matter.

"I know it puts you on the spot, but I thought maybe you could at least stroll by and, if you think there is nothing to it, I can reassure this lady," he concluded.

I agreed to do it and immediately left my own patients waiting and went over to our hospital. On the floor we used for pediatrics, as soon as I raised the question of any seriously sick children being cared for, one nurse excitedly whispered,

"Oh yes, Dr. Dailey, we don't know what to do. We've really been worried, and the child is definitely worse since coming in late last night." She went on to explain that the patient was one of Dr. Kelly's that had been admitted late last night. They had been giving intramuscular injections of penicillin as Dr. Kelly had ordered, but the child seemed to be worse, now not responding to its name and moaning whenever it was touched. She informed me that they had called Dr. Kelly first thing this morning, and he had told them that he would be out to check the patient, but he hadn't as of yet. She pointed out that it was close to noon and Dr. Kelly was expected any minute, but they had already decided if he were late or if he made no care changes, they were going to call one of us. She was more than happy to lead me to the patient's room where I slipped my hand beneath the head and neck of the child and experienced only tightening and rigidity as I tried to bob the child's chin toward its chest, suggestive of meningitis. We returned to the nursing station and formulated our plan. Since Dr. Kelly was now confirmed to be en route to the hospital, we could be diplomatic. Immediately upon arriving, the nurse was going to take Dr. Kelly aside and express her extreme concern over the child's condition, pointing out that it had worsened in the hours since admission. At this point she was to stray from the truth a bit and tell Dr. Kelly that I had been on the ward for other reasons and had noticed this child through the open door. The plan called for the nurse to tell Dr. Kelly that I had come into the nurses' station and casually mentioned,

"Boy, I don't know whose patient that child is in room 10, but it looks like it could have meningitis. If it were my patient, I'd be getting it on over to the center at Dalsville, probably to Dr. Quail if I could reach him." Back in my office, less than 30 minutes later, a phone call from the nurse confirmed the plan had gone perfectly, and the child was being loaded into the ambulance as she spoke.

My incident related, the administrator and each of my office associates took turns volunteering their opinions that it was necessary to revoke Dr. Kelly's hospital privileges. Dr. Zelle, although noted to nod affirmatively during these opinions, had remained otherwise silent. Finally, Dr. George asked him directly what his thoughts were. Put on the spot, Dr. Zelle drew in a long breath and with a nervous and pressured voice agreed that for the good of the patients, Dr. Kelly needed to give up his hospital privileges.

"We'll deal with it at the next medical staff meeting," Dr. Zelle pronounced. Realizing how difficult this would be on Dr. Kelly and also all the others who had practiced all their professional lives with him, I asked,

"How do we go about it? It's going to be tough, you know."

"We just do it," snapped Dr. Zelle. "That time is going to come for us all, and when it does for me I hope one of you will just tell me it's time to hang it up. I don't want to be left alone out of sympathy until I hurt a patient," he added.

"Since you have the only objective evidence, Dr. Dailey, you present the story you just related to us, for all the medical staff, at our next meeting, and the rest of us will back you up with the concerns and stories we've heard from the nurses recently," Dr. George suggested. The others nodded their agreement with this approach and one of them muttered,

"We'll just have to hang tough and get through it." Although anything but pleased with the role assigned to me, I found myself nodding affirmatively.

Hoping that Dr. Kelly might be absent the night of that meeting, but knowing it would only prolong the agony, I entered the meeting room that night to find him already there. He was leaning back in his chair with his eyes half-closed and looked relaxed and secure like he didn't have a care in the world. Why should he not appear at ease, other than with his family, Dr. Kelly felt more at home and secure with his colleagues of the medical staff than anywhere else he could be. The meeting began as usual with minimal old business being dispensed with in a matter of minutes, and Dr. George's voice then announcing,

"Under new business tonight, Dr. Dailey has an incident involving questionable care rendered by one of the medical staff that he wants to share with us." Literally, you could have heard a syringe needle drop in that room. It was so quiet that I could hear the labored emphysematous breathing of Dr. Zelle. I stumbled through the story as before, and this time I added that the child had been determined at the center to have had H. Flu meningitis but had responded well to therapy and was now back home and doing well. After a seemingly interminable silence, during which the only sound was that of Dr. Zelle's labored breathing, Dr. George spoke again asking,

"Well, how about the rest of you. Has anyone else observed or heard any complaints in regards to Dr. Kelly's care of patients?" After absolute silence from all for several seconds, Dr. George began addressing the question to each physician in-turn, around the table. All answers were the same, short and negative. With the beginnings of nausea I listened to the individual responses: "Not really," "No," "Nope," "I don't ever hear anything around here," and so on.

"Well then, Dr. Kelly, do you have any response to this story as Dr. Dailey presented it?" Dr. George asked. Again silence, as Dr. Kelly shifted his weight forward in his chair and removed the ever-present toothpick from the corner of his mouth before muttering,

"Well, ah hell, you fellows have got a man drowning before he knows he fell out of the canoe."

Dr. Kelly might have fallen out of the canoe that night, but I felt like the one drowning. I've reflected many times over the years how unfairly I was used during that meeting but always with the same gratifying conclusion. Those physicians met their professional responsibilities to their patients and to their hospital that night. They did not ignore the facts, look the other way, or protect their own. The profession was definitely still alive then. When they became aware of the signs of a problem they took action. Even though, that action for them was the equivalent of a family asking one of its own never to associate or come to the house again. They policed their own and without resorting to burdensome committee meetings and costly and paranoia generating paper trails for doing so.

The workload in the '70s was tremendous, and it was not uncommon to get home from the office at 7:00 p.m., only to be called out to the emergency room for another hour or two. We didn't actually perceive it as that stressful though, and this was to a large part due to the doctor friendly atmosphere of those times. This was an era when the hospital staff existed for serving patients and their physicians first and foremost. In those times, the hospital staff was not focused on serving The Joint Commission (a non governmental entity given the responsibility for periodic inspections of the nation's hospitals with regard to safety and quality and, which, has mysteriously become the prime player for generating policies and guidelines under which hospitals operate) and corporate management's profit-line before anything else. There was still a feeling then of camaraderie and common purpose among physicians, nurses, and hospital administration, nothing like the suspicion, dissention, and distrust that frequently exists between these groups now. Back then, most of us were totally trusting and dependent on our nurses, and we had excellent ones. This is not to say that there are not excellent nurses today, but they are rapidly becoming outnumbered by their less than adequate colleagues sharing the same title but having much less clinical science training and clinical experience.

In the '70s, nursing staff consisted primarily of licensed practical nurses (LPNs), three-year degreed registered nurses and four year B.S. degreed registered nurses. Each category had a medical science knowledge base distinctive to their group, and that knowledge base increased proportionately as you went from LPN to a four-year BS degree RN. If you encountered a nurse with an LPN ID tag on her person, you could be assured that this person possessed a minimal knowledge base of medical science. This was because in those times, the number of nursing schools was limited, most were associated with large urban hospitals, and they had uniform curricula which were well-grounded in the basic medical biological sciences. Every nurse then finished his/her schooling possessing a certain knowledge base in medical sciences which benefited patient care and also served as a common bond for understanding and interacting with physicians. Because of this emphasis on the medical sciences in their educations, nurses in those times were not likely to report by phone to the physician that the patient's 6:00 a.m. fasting blood sugar was 70 and then ask the doctor if he/she wished to withhold the morning's dose of intermediate acting insulin (A physician would not want to withhold this type of insulin since *intermediate* insulin is designed to act several hours *after* being administered, when the blood sugar *will* be high). I'm frequently asked this same telling-question under similar circumstances by nurses today. Similarly, nurses of that era, because of their basic science background in microbiology, were much more likely to wash their hands between patients and procedures than their counterparts today which, without sufficient medical science educations for understanding why hand washing is imperative, are simply taught,

"Before X procedure you always wash your hands, and don't forget to make this a protocol and keep a copy of it in the nursing policy manual to prove that you are doing it and so that you can train all new nurses with it."

Frequently today I see nurses with ink pens in-hand, leafing through an opened chart, periodically laying down the pen and reaching into a bag of chips for popping a chip directly into their mouths. This tells me all I need to know of such individuals' basic medical science educations in microbiology. Should we be surprised that hospital acquired infections are killing thousands of patients annually?

The relative emphasis on the medical sciences in the educational programs of nurses of the sixties and seventies led to nurses back then not being on the phone to their physician friends during the flu outbreak, asking for an antibiotic prescription for their families, as their education was sufficient for knowing that antibiotics would be useless against the virus causing the flu syndrome. Commonly today, however, I hear this request from nurses. In times past, nurses had better formal educations in the medical sciences, enabling them to progress professionally much more easily and quickly. Past generations of nurses assigned to specialty nursing areas, such as the CCU or the neonatology units, were much less likely to mistake their in-depth and marked skills within these focal areas in the hospital as the equivalent of an MD's in-depth and overall knowledge base in all areas. This was because nursing educational programs in those days provided background in the basic sciences sufficient for enabling them to recognize and appreciate how little the medical-knowledge surface had been scratched by their education, compared with a physician's knowledge gained from ten or more years in a traditional medical school and residency specialty education.

Unfortunately, it was also in these earlier years that I noticed the first crack in the traditional pathway to the RN degree that was a harbinger of things to come. We had in our community a small college which had two sister institutions, one in another southern Illinois town and the third in a southern Indiana community. Ours had a relatively new LPN program associated with it, and if judged by the number of local applicants, it was a very popular program. Naturally, our small hospital took advantage of this local supply, and within a very short time the preponderance of LPNs utilized by our hospital were graduates from our own local community college. These skewed numbers came to light primarily as the result of the physicians' complaining that it seemed the quality of our LPN's was decreasing. Specifically, they were felt to be lacking in basic medical background and, particularly, in clinical skills. In short, most of the newly hired LPNs not only lacked book and classroom knowledge in medical sciences but had never even seen many common clinical procedures performed, let alone having had directly performed any themselves. About this same time frame, I was contacted by one of our community college's sister facilities as to whether I would be interested in serving on a nursing school advisory board for overseeing an expansion of the LPN programs of the three institutions. Young, full of my self, and still naïve enough to think I was being asked because they really wished to take advantage of my medical education and experiences, I jumped at the offer. My tenure with this board was extremely short, and after the second meeting I tendered my resignation, sighting incompatible philosophies as my reason. In the two meetings I attended it became clear that their definition of the expansion of the LPN program was to begin calling themselves a two-year associate <u>RN</u> program, while minimally and insufficiently upgrading their academic staff or opportunities for clinical experiences above what they were currently providing for their LPN programs. It seemed obvious to me that the motivation behind this plan was not the production of well trained RN's. The shortage of RNs, an already established but struggling community college concept, and pressure from the business corporations now managing our hospitals with expansion and profit as their goals, provided the ideal atmosphere for the community colleges to increase their enrollment. Even after explaining to the board, that from a physician's perspective, their current LPN programs were producing inadequately trained LPNs for employment, they could not satisfy me as to how they could justify attempting the education and clinical training sufficient for a two-year RN degree, especially given the increased responsibilities in clinical care which they intended bestowing on these 2-year RNs in comparison to those permitted the LPNs.

Years later, reflecting on this experience, I think that I was privy to see the future for RN education and the beginning of the end for consistency in knowledge and clinical skills among newly graduated <u>RNs</u>! As in so many other areas in medicine, the outside forces of economy, entrepreneurialismism, and human nature had taken its toll on the nursing profession. Fueled by the media's constant cry of nurse shortages and the guaranteed high salaries of the taking-off health management corporation era, people flocked to the two-year RN programs. After all, if your motivation for attending school was a professional title with good pay, why would anyone choose a four-year route or for that matter even a tough, *high-standard,* two-year program. In short, people were offered an over-paid professional title for only two year's academic investment. What an incredible bargain!! The two-year program door that opened at that time has swung even wider with passing years, as more and more smaller colleges and community colleges jumped on the band wagon in competition for enrollment. The result has been far reaching and significant for physicians, hospitals, patients, and most of all, nurses themselves. For the vast majority of the great numbers of these newly milled and stamped nursing graduates, only two major benefits accrue as a result of their two year curriculum. If fortunate, their two years educational investment enables them to pass the state boards, since one of the mechanisms of teaching in the two year programs is, I am told, focusing on material known to be on state board exams. The other major benefit from the two year program is the coveted <u>RN degree</u> (albeit only a two-year associate one) which opens the door for lucrative hospital employment and, for those motivated enough, an on-the-job real nursing education over the next few years during their employment. Indeed, a few of them eventually become quite competent and a few even excellent, but only after several additional years of self-teaching and learning on the job, sometimes at the expense of medical staffs and patients.

At the time I served on the board of advisors for our community college, the two-year associate nursing degree concept was in its early infancy and more than a few within the community college faculty disagreed with my views and predictions regarding it. Within a year of the initiation of their new RN program my own office nurse, who had a B.S. degree, surprised me with the fact that her husband had taken an out of state job, and she would be joining him within two weeks. One of my associates came up with the idea of calling our community college and seeing if they had a student in his or her clinical phase that might benefit from real clinical experience while drawing a salary for two weeks, thereby giving me time to search for a permanent replacement. A phone call to the director of the program revealed that she thought the plan excellent for both parties, a win/win situation as it were. She assured me she had an ideal student in mind, one of their "better students" who had come up through their LPN program and then continued into their two year RN program where she was now in her second year and beginning her clinicals. The following day I was in my office when my nurse opened the door smirking and announced,

"My replacement is here." I immediately asked her what she thought of her. "You'll have to decide that. I really don't know anything about her," she said. But her smirk was mischievous and I suspected trouble. Entering the room where my nurse had left our new applicant, I immediately had one of those uneasy feelings about the young girl seated before me. It was like I had known her previously, and the experience hadn't been positive. After several minutes of chatting with her I still could not place her or recall the circumstances, but I was certain I knew this person from somewhere. Her interview was satisfactory and we agreed that she should start the following day, and when she had gone, I rushed to my nurse and blurted out,

"Alright, enough smirking, now tell me where I know her from," I demanded.

"She and her kids were regular patients here until two years ago," she laughed.

"Well, that's not necessarily bad", I replied, but quickly added, "Unless I honked them off or something." Giggling now, she teased,

"I'll give you a clue. She had a small baby and a young child and they always were in need of a good bath."

"So what, a lot of single working mothers have periods in their lives where it is tough to meet every responsibility exactly on time. Besides, her hygiene appeared good today," I defended. Chortling now she quipped,

"Last clue, spaghetti was stuck to her foot," she announced before decompensating entirely in laughter. I couldn't believe it, but she was right. I forgot the person but not the scene. This was the lady that had come to us a couple of years back for her yearly pelvic exam and left alone by the nurse for undressing and putting on her gown. A few minutes later my nurse and I entered the exam room to find the patient already on the exam table with feet in the stirrups and lower body and feet draped with the sheet. Positioned on my stool at the end of the table, I pushed back the sheet and found myself seated between two stirruped feet, the soles of which looked like they had been barefoot in a newly turned garden all day. Eyes riveted to them, I found myself instinctively leaning away from the left one when I saw what appeared to be a large worm adherent to its sole. Motioning my nurse to my aid, I silently pointed out the situation. We both looked again and then at each other silently mouthing simultaneously, "Dried spaghetti!"

Undaunted and unprejudiced by these memories of the lady, we launched into the week's routine, my temporary nurse and I, both in good spirits. By the end of her first week's duty it would have been truthful and fair to say that she presented a good appearance, was pleasant enough, but fell considerably short of even mediocrity in her nursing knowledge base. Just before leaving for the day at the end of her first week, she approached me for advice on what she should do for one of her children. She described for me what sounded like impetigo secondary to scratching and introducing infection to insect bites of the child's ankle area. Wishing to save her any expense that I could, I offered her a small carton containing several tiny unit-dose sample bottles of an appropriate antibiotic. Each miniature container held a small amount of the antibiotic powder and required about a teaspoon of water for reconstituting the dose. I showed her the amount of water needed and cautioned her to shake it well before giving it to her child, pointing out that it should be used four times a day for seven days. She thanked me, told me good night and out the door she went. Home about an hour, I answered the phone to recognize her voice on the line,

"Oh, Dr. Dailey, I forgot to ask you. Do you rub that antibiotic on the infected bites, or is he supposed to take it?" I successfully resisted ever commenting on this episode to our community college's nursing director but at great sacrifice by the "I told you so side of me".

These were full and hectic days. None of the three of us in our group took appointments. We operated on a sign-up and wait your turn basis. Driving by our building at 8:00 a.m. before we were open, it looked like a public auction with people milling about waiting for the doors to open and first chance at a seat in the waiting room. Once you began seeing patients it was non-stop until all had been accommodated. Lunch hours when they came at all, were frequently lunch minutes, and 7:00 p.m. closings were not unusual. My basic office call fee in those days was a whopping $7.00, and when I hear younger doctors today speak groaningly of 30 patients per day office days, I think that back in those $7.00 office charge days, we were like the old popular investment firm ad. We made money the old fashioned way. We earned it.

As today, the main complaint from patients then was short contact time with the physician, along with having to wait to be seen. Periodically I would have to defend myself against one or the other of these two complaints. Most often it would begin with an irritated patient's suggestion,

"You know doc, if you would go by appointments we wouldn't have to sit out there so long." In reality, a primary care physician's office attempting to schedule appointments frequently results in even more rancor among the patients. Usually I was able to make them understand this by presenting them with a little pretend scenario.

"Suppose I am seeing Mrs. Smith for her scheduled 10:30 a.m. appointment. She's walking down the hallway to the exam room, and you come in the front door with your freshly sliced finger dripping blood from the handkerchief you've wrapped around it. Would you want me to honor my commitment to Mrs. Smith's appointment or to attend your wound instead," I would challenge. In reality this is exactly what used to keep most primary care physicians running behind schedule. Unscheduled trauma, cardiovascular events, and a host of other potential schedule disruptors were always lurking in the wings. Sometimes a patient would hit me with,

"Whoa, doc. You're in and out of here in a flash. I had something else I meant to ask you." Often in this situation the patient would frame their complaint by pointing out that he or someone he knew went to doctor so-and-so in the center over in the city and that they scheduled half hour appointment slots over there. In such instances my defense was educational and easily mounted by simply asking,

"How much was the office charge there?" If the difference between ten minutes and thirty minutes with the doctor was noticeable, the difference between $7.00 and $20.00 for a visit was certainly so.

We were aware of locals who selected a personal physician from our nearest metropolitan center rather than seeing a local doctor. Doctors have feelings too, and at the best case scenario, we were able to keep them from being hurt beyond recovery by rationalizing that such patients chose to go to the city for their care because our hospital was old and perceived by them to be outdated. The other much more unpalatable possibility was, of course, that they did not think any of the local doctors were as good as the physicians in the city. Although difficult to accept, at times it led to some enlightening discussions and amusing situations. As products of U.S. medical schools, we knew that U.S. medical schools' curricula are standardized and equivalent. In the World Health Organization's ranking of medical schools the U.S. schools are near the very top. There is no such thing as a student transferring from one U.S. medical school to another U.S. medical school because the other is easier or less demanding. Similarly in those days, no U.S. medical school had to fill its seats with less academically accomplished students because some other school had gotten all the good candidates. We could not expect the general public to be aware of this fact since the small town's other personal-need profession for comparison was that of attorneys. The professional reputations of law schools seem to run the gamut: the disreputable, the mediocre, the good, the excellent and the best. Whenever the opportunity presented itself I would try to educate patients about the homogeneous continuity peculiar to the U.S. medical schools and remind them that doctors are people, and some people not liking urban settings, opt for lives in the rural communities. This was an entirely truthful statement in those days since the physician glut of the early nineties that eventually forced many physicians to locate or relocate within small towns had not yet occurred. A favorite way of mine for getting this point across was to use the vocation of the person I was in discussion with and bait them with,

"What about you, Joe? You're an accountant, so why are you living here? Are they all filled up with accountants over in the city, or did you just think you didn't know enough to be an accountant in the city?"

I certainly wasn't immune to hurt feelings caused by patients who preferred driving the 35 miles to the much larger Dalsville for their medical needs. One of my most memorable patients for demonstrating the public's lack of knowledge regarding the skills and training of various members of the medical profession, came to me via my volunteer work at the county well-baby immunization clinic. Initially the clinic took about two hours of my time one day a month. Its purpose was to provide indigent and Medicaid patients a place to have their children immunized at a cost they could afford. My job was to check children over for assuring no contraindication existed to their receiving immunizations. Also, I was to provide answers for the parents' questions regarding their children's health and to make suggestions for necessary follow-up of any problems that might come to light during the exam. In the incipient months of the clinic we saw between ten and fifteen children per clinic, and I treated the time away from the office as a long lunch hour. By the end of the first year, however, the number of children had risen to thirty, and I was not only donating my time but one-half day's lost office revenues as well. In addition, the well baby clinic had melded into a treatment clinic. Since the majority of the clinic families had little money and no insurance, often I would make use of sample medications. If the family already had a local doctor, and if the child was found to have an ear infection, for example, we would refer them to their personal physician. This worked well for the indigent families then, as in those days your personal physician carried most of these cases on his books and utilized sample medications as freely as possible. For all the undoctored patients we used a sample medication, or I made an appointment with my own office, knowing full well that I would not be paid for seeing them. On the rare instances where a problem was found requiring specialty consultation, my pediatric colleague in Dalsville whom I had looked into Dr. Kelly's meningitis case for, was kind enough to take the referrals. By the end of the second year of these clinics, I found myself dreading them due to their size and time requirements, and I asked the county health department to please see if they could cut down on the numbers involved. Their usual response was to remind me how greatly these families appreciated the clinic and what it meant to them to have access to a doctor. In fact, the nurses pointed out that this was precisely why the clinic numbers had risen. It was a service these families needed. Near the end of one particular clinic, tired and irritated by the larger than usual numbers of that day, I examined a two month-old who had a very bad case of diaper dermatitis. I explained to its young mother the things she could do to prevent its reoccurrence in the future and ended with telling her that the condition was bad enough now that it would require prescription cream for clearing it up. She hesitated a few seconds, and I thought the prospect of having to pay for a prescription was being processed in her mind so I added

"If you like, when this clinic is finished, if you will ring the doorbell on the back door of my office my nurse will bring you out some samples of the cream you need." Already dressing the baby and gathering their things together, she replied,

"Oh no, that's alright. We'll just drive on over to Dalsville and see our pediatrician." Totally caught off guard by the rejection of my offer, all I could splutter out was,

"You have a pediatrician in Dalsville that the baby sees?" Knowing full well that even if this baby had a Medicaid card from our state, Dalsville was across the state line and that meant the card would not be accepted there. "Isn't that a little expensive and a long drive, considering you can take the samples and clear the rash in a few days?" I asked. Baby already over her shoulder and headed for the door, she delivered her final blow to me.

"The drive is not bad, and we can scrape the office-call money together if it means getting our baby *the best medical care.*"

Dumbfounded, I went home to lick my emotional wounds and analyze the significance of what had transpired over a case of diaper dermatitis. Angered and hurt I reflected on why, eventually asking myself the question, was what just happened to me any different from rejection and insult in any other vocation? On the most basic level, no, we all have pride, and no one likes rejection. I had to admit, however, that physicians as a group seemed much more vulnerable to rejection than other vocations and professions. Once hit on, the explanation for this was incredibly simple and consisted of the increased pride inherent among physicians. The rigors and nature of the extreme educational requirements for physicians are the basis of this. Indeed, pride itself is a necessary tool of survival for pre-medical and medical students. Without it, they could never withstand the extreme, harsh and lengthy educational process. Actually pride is already a characteristic inherently common in those deciding to pursue a medical education. By nature of the beast it is, to a large degree, what directs these folks to choose medicine as their career to begin with. As undergraduates, first year pre-medical students had more pride than their freshman non-premed classmates. The intense competition for high grades in academically challenging chemistry and biology courses during these years requires an excess of pride for success. When pre-med students are sitting in a library cubicle on a Friday night while the conference rival basketball team is battling it out in the arena three blocks away, it's their pride that sustains them and keeps them focused on their goals. By the end of these pre-medical years, for those successful, the process and experience results in an even greater sense of pride, and thank goodness for that fact, if medical school is to be survived. No person, regardless of educational level, has even an inkling of what it takes to survive four years of traditional medical school in the United States. Only a doctor who has done it or a spouse who survived it with him or her can possibly know. A physician friend of mine, reflecting on threats he perceived in the wings for medicine as a profession, once commented to me,

"You know, however bad it becomes in the future, and even if the profession is lost entirely to special interests and big business, to have successfully competed for a seat in medical school and then to have stayed the course with those fellow academic cutthroats is the greatest fraternity to belong to in the entire world! Nothing can take that away from us." How descriptive and, unfortunately, prophetic his comment was. It is then, this increased personal pride that makes physicians so vulnerable to hurt feelings when rejected. They enter their practices with pride levels greater than all other professions and vocations. The public often mistakenly refers to this as big ego, but that is incorrect.

Ego implies an inflated pride without any substantiating basis, but physicians, by virtue of their extreme educational requirements, maintain a well-deserved pride, not inflated egos.

Although I could not expect that mother in the well-baby clinic to have any concept of physician educational requirements or pride levels within the profession, there was something beyond this issue that continued bothering me about the incident. Over the next several days I repeatedly revisited the event and sought to explain what still annoyed and worried me about it. When it finally came to me the answer was both simple and complex. The simple portion was the ludicrous misuse of the medical system hierarchy by the mother having insisted on a pediatrician for caring for a case of diaper rash. And the complex part was the educational, cultural, psychological, and media forces that contributed to that mother's misperception of what was needed for her child with diaper dermatitis. From my perspective as a family practice physician, I felt extraordinarily over-educated for treating diaper dermatitis, and yet this mother felt it was necessary to take the case to a physician of even more focused training. It was a terrible misuse of resources back then, and even today with plenty of all types of physicians and medical caregivers, would represent an economically inefficient use of the system. Retrospectively, that mother and child were harbingers of what was yet to come in the escalation of the cost of medical care. Specifically, they were the forerunners of the public's misperception and fascination with "specialists" which would eventually, with the help of our medical schools and even more help from the media, become what I refer to as the anatomical specialty mindset. As a result of this mindset, if I were dealing with the diaper dermatitis mother today, I would still be terribly over-educated for treating diaper dermatitis and the Dalsville pediatrician even more so. But it wouldn't matter today, because that same mother would likely opt for a dermatologist!

With a better understanding of the diaper dermatitis event, my feelings were soothed somewhat but my stamina remained in jeopardy. Thus at the next clinic, we reviewed the files of our patients and found astoundingly, that nearly fifteen percent of our regulars did see pediatricians or family doctors in Dalsville. This meant that they were able to pay cash for an office visit there, while attending and swelling the ranks of our local immunization clinic for free. Finally I was able to reduce the size of our clinic a little, about fifteen percent, to be precise!

A more amusing episode that grew out of my association with children and the well-baby clinic, which also demonstrated the public's lack of sensitivity and knowledge regarding medical resources, pride, and the referral system, involved another child whose personal physician was a pediatrician in Dalsville. It was about 6:00 p.m. on a Sunday afternoon on the backside of one of those hellish weekends on call for the three of us. I had just gotten home from the ER without a beep en route or a message waiting, and it was looking like I might get to eat the Colonel Sanders I had picked up on the way. The rest of the family was away for the weekend and so, as long as this quiet spell might last, I could enjoy my food and watch whatever I wanted on TV until the next barrage of calls began. At least these were my thoughts as I flipped the TV on and began transferring chicken anatomy to a plate, only to be interrupted by the ringing phone before even my first bite. The voice on the line was female, young and began apologetically,

"Dr. Dailey, you don't know me or my baby because I see another doctor here myself and my baby has a pediatrician in Dalsville."

"Oh," I replied during the pause, not knowing what else to say and sensing she was going to add to this.

"Well, we've heard that you're good with kids and babies, and our baby has a temperature of 102 and we wondered if you would see it," she finished.

"Have you called the baby's pediatrician in Dalsville?" I asked.

"Oh yes, we did that this afternoon when the temperature first started," she assured me.

"Well, what did he say?" I countered.

"He said it sounded like a viral infection, and he hated for us to have to drive all the way over there if that's all it is. He said that there's no way to be sure without actually having the baby looked at," she explained.

"So, why not just take the baby on over and reassure yourself?" I suggested.

"Well, the pediatrician said if we brought the baby over it would be screened in the emergency room there, and if it were sick enough for admission or anything complicated, then he would come in to see it," she explained.

"So, I don't understand. Why are you calling me then?" I asked.

"Well, the pediatrician said we might want to just take the baby to our ER here instead of driving all the way over there, but we told him we didn't like any of the doctors locally, and that is when he recommended you," she finished in an upscale tone, obviously glad for getting something out she thought would please me.

"Well, I am pleased that your pediatrician spoke well of me, but I am not the doctor on call to the ER tonight. I'm on call only for our group and I've been in and out and on the phone for most of the weekend, and I just got back in for the evening," I explained.

"You mean you won't see us in the ER?" she asked, her voice becoming more urgent.

"Well, considering you have already been offered options by the baby's own doctor, it's not like you don't have other choices," I tried to diplomatically explain.

"I don't believe it. My baby's temperature is 102 and you're a doctor and you won't come out and see it!" she screamed.

"That's just it, I'm a doctor, but I'm not your baby's doctor, and I'm not the doctor on-call to our ER. I'm dead tired from a weekend on-call for our patients, and you have other choices. It's not like I'm the only doctor in the area," I reasoned with her. Losing control now, she hysterically shrieked,

"It's late now and by the time we could get ready, drive to Dalsville, and wait to be seen in the ER, it would be midnight before we got home! I can't believe that our pediatrician gave us your name and you won't see us," her shout trailed off into tears. "Here's my husband. He wants to talk to you," she blubbered.

"What do you mean you won't see our kid? What kind of doctor are you?" a husky, male voice blasted at me in a macho and threatening tone. I went over again the explanation just given his wife, interrupted periodically by his interspersed, "Well, shits." and "Well, hells." When I finally finished my explanation again, there were a few seconds of silence followed by his voice again, this time in a lower, more menacing timbre.

"Look," he began. "I know you haven't lived here long enough, and you might not know me. My name is Bill Barney and I'm known around the area as a man you want no part of! I've whipped a lot of ass around here, and I know right where you live!" He then proceeded to describe my house in detail and the street names whose corner it was located on. It was now my turn for a few seconds of silence so he added, "I just might come on over there and make you change your mind."

"Well it would be a good night for you to do that since I'm the only one here," I shot back, and then added, "but you know what, even if you come over and beat me half-to-death, I still won't see your baby tonight. And by the way since you know so much about me as to know where I live, you might also want to know that I'm a marksman and collector of pistols..." He interrupted me here, shouting,

"Yeah, now that I've given you warning you'd be ready for me and shoot me, wouldn't you?"

"The very minute you step onto the porch," I bluffed.

"Well I'll just wait and catch you out one of these days, maybe late at night when you're coming back from the hospital," he threatened as I was hanging up on him. The next day I asked Dr. George about Mr. Barney, since Dr. George had reared three sons in the community and was well known to have engaged in some fist-to-cuffs himself when the occasion called for it. Dr. George must have been lucky. According to him he had never had his "ass whipped" by Mr. Barney and did not know the name either.

Dr. George was a good person to go to if you were interested in toughs and hooligans in the community. He had practiced there since finishing his medical training, and he was opinionated and outspoken. This, along with his years spent on the school board, hospital staff and part owner of the local newspaper, had provided him with plenty of exposure to such individuals. He had played college football, served as a WWII pilot, and the little leisure time he had was spent carrying out physical labor on his cattle farm. At fifty-three, he physically was still intact at 6'-2" and sported a very muscular frame. His patients all dearly loved him, and the remainder of the community either liked him or hated him. There was no middle of the road feelings towards Dr. George. Growing up, all three of his sons had been outstanding high school athletes in basketball and football, and Levelton was first and foremost, a football town. Even though his sons were long graduated from Levelton High, Dr. George had remained an avid fan and athletic supporter, frequently serving as team physician at home events. His largest contribution to the program, however, was his annual pre-athletic exam day. Our office would shut down for one-half day, and the three of us and our nurses would go to the high school's gym where any student planning on participation in any of the school's athletic programs from grades 7 through 12, could be screened and examined free of charge if they chose. Needless to say, this involved a large number of students, and whether viewed from the extra work perspective or lost office revenues, represented a very significant donation to the school system and community. One fall, a week or so after the athletic exams had been carried out, a young man presented himself at my office as a patient, registering and waiting his turn like the rest. Entering the exam room he quickly admitted that he was not there for medical advice but only wanted a few minutes for introducing himself and telling me of a program he was interested in. His name was Charles, and he had come to Levelton from Cincinnati, where he had been educated, trained, and certified as an athletic trainer, even serving with the Cincinnati Bengals for a period of time. He had recently been hired in sales with one of our local companies and was thrilled to learn that Levelton had such an outstanding football program. His hope was to be able to volunteer his services as a professional trainer to that program, and he wanted to, as he put it, "touch base" with some of the local physicians to see how we felt about his plan and to hopefully obtain, in his own words, "some backup in the medical community" which he might call on when needed. I assured him that I agreed with him generally, and anything that would improve and benefit the program would be good for the community in my estimation. Thanking me for my time, he left, assuring me that I would be hearing something more definite around the start of football practice. With that innocuous introduction began one of the most frustrating patient/doctor public relations confrontations I've ever known of. It would span three years time, with acrimonious public meetings, uncomfortable confrontations, and eventually, it would drive a wedge between the entire medical community and the school system.

Charles never did get back to me at the beginning of football practice as he had promised. Instead, my next contact with him came after our first football game of the season. The ER called my home around 11:00 p.m. and notified me that Charles was there with one of the players with an injured shoulder, and he was wishing it to be x-rayed. It happened that the player was normally one of my patients, so I gave the okay for the x-ray and told the nurse if she or Charles needed me, she could give me a call. If nothing more was needed, I instructed her to have the patient come into the office for follow-up the next day. The patient did show up the following day, and an exam and review of the x-ray showed no fracture but suggested a mild acromioclavicular strain. I gave the patient instructions along with a prescription for anti-inflammatory medication and was about to advise him on his limitations, when he interrupted me with, "Oh, Mr. Charles is going to take care of all that. The coach said to tell you that." Somewhat surprised at how entrenched our new athletic trainer had become in such a short time, I mentioned the episode to my two associates and then forgot about it. The weekend came and several of us were having dinner at the local Elks Club, and there was a table of local businessmen that probably was only within earshot due to the number of their *shots*, but this enabled me to hear that the topic of their discussion was the opening football game earlier in the week. It quickly turned to the new athletic trainer and what a plus it was for the program to have someone of his expertise on the bench each game. This set me to reflecting again how successfully and quickly Charles had sold himself to the community, as well as the coaches, but I stopped short at the sound of my own name followed by, "Yeah, Charles said that the new young doctor here didn't even recognize that the clavicle was fractured." The remaining food on my plate tasted like cardboard as I hurried it down in order to get out of there and check our radiologist's report on that clavicle. Back in my office at 9:00 p.m. on a Friday night, my anxiety quickly gave over to anger as I read over the radiologist's report of the clavicle in question which read, "No evidence of any fractures noted on this study." As bad as that particular incident with Charles seemed that night, it paled in comparison to the injustices that were yet to come over the next two years, not just to the local medical community but to patients as well. It was this last point that was most frustrating and frightening of all, since the patients and their parents appeared to be eating up those injustices and standing in line for second helpings. The attitudes of those parents demonstrated the complexity of our medical system and how totally unequipped the general public is when left to their own devices for making safe, effective and economic use of it and how vulnerable patients can be. It would become even more evident in the future but was already being demonstrated back then, by the bizarre experience with this particular athletic trainer and his influence with well-meaning but gullible parents, that the only person who is really capable of utilizing our complex medical system in a safe, efficient, and economic manner is a physician. During Charles's first year as Levelton's football trainer, the signs and signals of his more personal game presented themselves fast and furiously. None of the three of us in our office saw any athletic injuries that season, and our colleagues in the community confirmed the same trend for their practices. As spectators at games and in following the local newspaper's sports coverage, it was obvious that the usual injuries were occurring and that another physician's name began to be mentioned frequently as well. This doctor was an orthopedist in Dalsville, and I was vaguely aware of him, as he had begun his practice there about the time that I was finishing my internship in the center's hospital there. By the end of the first football season with our new athletic trainer on the bench, it was clear to us all that any injury requiring more than first aid was being sent to Dr. Nelson, the orthopedist in Dalsville. At this point it became a matter of principle and pride for the local physicians. After all, Charles had been less than truthful on his one-time visit to us. Dr. George, having a long history with the school system and particularly with its athletic programs, was elected to speak with the high school principal and head coach about the situation. In retrospect, this may have been a mistake, in that the principal and the coach apparently looked at Dr. George's concern as one of losing his long standing voluntary role of athletic consultant. As soon as this implication was made, Dr. George in his inimitable manner took offense, and soon the meeting was rancorous with phrases such as "loss of business", "GP's", "specialists", etc., etc. The one illuminating point garnered from that meeting was that the head football coach had himself undergone disc surgery by Dr. Nelson the summer prior to the football season.

This prompted me to place a call to Charles and quiz him about the trend we were observing. His story was that he had learned of Dr. Nelson's presence in Dalsville and of his alleged special interest in sports medicine, and so he had begun referring cases to him. He assured me that Dr. Nelson had special training and a special interest in athletic injuries, and that he was well-known for this in the Dalsville area. I informed Dr. George of this latest turn of events, and we decided additional intervention was needed. We arranged a called-meeting of our local medical staff, making sure that our pathologist and radiologist were present. We felt that the two of them, being from Dalsville, would make very credible corroborating witnesses, especially since by this time both were more outspoken on the matter after finding out that Dr. Nelson had recently lost his privileges at one Dalsville hospital and was dangerously close to losing them at a second. Our plan was to invite the head coach of the football team and the high school principal and to present our latest evidence and concerns. We did exactly that, and despite the physicians from Dalsville testifying that Dr. Nelson had been removed from one hospital staff in that city already, we gained no ground. The mindset of the principal and the coach remained that we resented the intrusion onto our turf and were willing to deprive local athletes of access to a "specialist" in order to defend our own business interests!!

The day following that meeting, my secretary approached me in my private office and with a knowing and anxious expression on her face, informed me in a whisper,

"Dr. Nelson is on the phone for you."

I lifted the receiver and in a self-confident and firm voice, which belied my actual feelings, I answered.

"This is Dr. Dailey. What can I do for you?" With a slight chuckling I heard his reply,

"Well, for one thing, you could stop bad-mouthing me to all my patients." I said nothing, so he continued. "Seriously, if there is a problem I need to know about then I'd like to hear it from you." I quickly summarized our concerns, choosing to concentrate on the questionable referral pattern of Charles and for now, at least, avoid the many questions regarding Dr. Nelson's professional ethics or his liberal surgical recommendations. Finished, I was shocked at his response.

"Well, Dr. Dailey, there is a very good reason why Charles refers so much to me. It simply is a fact that I am the best trained orthopedist in athletic injuries in Dalsville. He knows that. I don't see why you can't understand that." Bravado or ego, I couldn't believe this guy. "Well," he pressed me.

"Well, I guess for me it's analogous to a new drug becoming available. I don't want to be the last to take advantage of it, nor do I want to be the first to jump on board. I like to give it enough market experience that I can hear what my colleagues are saying about it," I explained to him. The not too subtle reference to what his colleagues might say about him gave him his turn for a few seconds of dead-line time and then,

"Well, you're entitled to your own opinion, but while you are over there crusading against me and running me down to my own patients, you might just find yourself with a slander suit on your hands," and with that threat he hung up. There was nothing subtle about that. I sat there a couple of minutes feeling anger, doubt, and more than a little anxiety that he might make good on his threat. Not all my anger was directed at him though. I was more than a little upset at the attitudes of the patients themselves, for not understanding the logistics of utilizing our complex medical system to their best overall interests and for not recognizing that this was one of the most important reasons for having a personal primary care physician. I resented their lack of trust in their town's medical community, and I was upset that they thought the medical system so simple as to be best accessed on the basis of "specialists" or whatever some physicians billed themselves as. I was most angered at them for not realizing the worth and training of their own primary care physicians, especially since credible estimates often quoted then were that in the neighborhood of ninety percent of all a person's health care needs could be competently met by well trained primary care physicians. Most of all, I did not understand how they could not appreciate how one of the most valuable services provided them by their primary care physician was guidance for safely, effectively, and economically accessing our complex medical system. What I eventually had to answer for myself was the question why I should continue to try to help people who clearly didn't think help was needed, and why I should continue attempting to save people from bad medical care who clearly didn't want my help and who also questioned my motives. The answer was miserably simple and the same after this far into battle as it was at its beginning. The medical profession then, still looked out for its patients' best interests and maintained the profession's integrity. The latter meant not condoning members who acted unprofessionally or in a manner detrimental to their patients. By this time, the matter was no longer limited to the medical staff, school officials, and patients. It was the topic of conversations in restaurants, homes, and any gathering of people, particularly at the athletic events. It had escalated into a full-scale public relations battle, and the victory spoils would be the trust and respect of our patients. Clearly, I had to continue with the battle but in a more cautious fashion, given Dr. Nelson's threatening attitude. I needed further corroborating evidence that it wasn't just the Levelton doctors who felt this way toward Dr. Nelson, but his own colleagues in Dalsville as well. Wishing to avoid calling Dalsville doctors and polling them, lest it get back to Dr. Nelson that I was spearheading a movement against him, I devised a slightly devious but effective plan for obtaining the evidence needed. I drafted a letter explaining that I had been asked by our local school board to recommend an individual orthopedic surgeon or an orthopedic surgical group in the Dalsville area that we could feel comfortable referring those few athletic injuries from our school system that warranted orthopedic expertise. I then asked for a list of three that they felt confident in recommending to me. I sent a copy of the letter to two family practice physicians, two general internists, and two general surgeons, all of them practicing in Dalsville. When the responses were all returned, it could not have been more supportive of my suspicions. Basically, there were two names that invariably showed up in the recommended three. The third spot on the list varied more often but included any of four additional names from the orthopedic community at Dalsville. In short, four names showed up frequently on the recommendation list and two names occasionally. The name that did not show up, not even once, was that of Dr. Nelson!! How about that as a forerunner of peer review! Unfortunately, this was my last major victory in this battle, as I was never able to change a single patient's, family's or school official's mind, even with the ammunition provided in those letters. Another season passed with yet more signs of overused orthopedic consultations and more signs of questionable diagnostic and surgical interventions. Three years later, after having returned to my home state and a new practice, one of the nurses at Levelton's hospital sent me the front page of the Dalsville newspaper on which was a photograph showing a parade. There appeared to be thirty or forty people, most carrying placards, and in the background was a large Dalsville hospital. The caption beneath it read simply, "Doctor Nelson's patients march in protest of his pending loss of hospital privileges." An accompanying article suggested he was moving his practice to another town. No, thankfully, it was two states distant and not Levelton! This was my first experience pointing out how vulnerable even a well-educated public can be when it comes to matters of healthcare decisions. It also was my first opportunity for seeing the power inherent in trust and for realizing that trust can be a double edged sword, particularly in medicine. These were educated and decent folks who put their trust in a doctor because they thought they needed a "specialist" for their kids to have the best possible medical care. Unfortunately, theirs was an unquestioning and blind faith, and they had demonstrated little knowledge of how the medical system operates most efficiently and effectively for the patient. They also underestimated the integrity, pride, and honor within a still extant *medical profession* at that time, and they especially underestimated the quality and ethics of the physicians of Levelton. Because of this they made the wrong decision where efficiency, convenience, economy, and saddest of all, in this unusual instance, quality of care rendered their children was concerned.

Levelton's hospital had not been much help in our struggle with Dr. Nelson for our own patients' trust. Deep in our hearts we knew this. It was a drab, yellow-brick, three-storied building of the Hill-Burton era and did little to suggest up-to-date quality medicine from its appearance. Just out of training, there were additional hospital services I too would have liked available, but by and large, I thought the staff was good and likewise the quality of care that the patients received there. On occasion I would have first-time patients come to the office and tell me they would like to have their records transferred and become established patients in my practice. Sheepishly then, they would proceed to the "but" part of their request which usually went something like, "but if I need to be in a hospital I would prefer to be in a Dalsville hospital. It's not that I don't trust you, doctor. I wouldn't be here if I didn't, but I just don't care for the Levelton hospital." Usually their opinion was not based on any past personal experience at our hospital, but rather on hearsay and always, the hospital's outdated appearance. I would reassure them that I would work with them in regard to their request and then use the opportunity for teaching. I'd begin by admitting that if I had no first hand knowledge of the staff or capabilities of our hospital, I might feel just as they did, based only on its external appearance. I would suggest to them that if the hospital were bad, I would not have chosen Levelton to live in, let alone to practice medicine in, since I had a family that might be using the hospital too. I would continue by explaining to them that there were numerous medical needs that our hospital could never meet and never should be expected to. For these type needs, referral to Dalsville hospitals would always be indicated. On the other hand, there were just as many needs that could be met every bit as well by our hospital and our staff, as they could by any Dalsville facility. Additionally, I would explain to them what I referred to as the home town advantage. "Don't you imagine that if you were my neighbor in Levelton, taught my children in school, or played bridge with my wife, human nature as it is, I might be a bit more focused on you than some patient that was passing through town that I didn't know from Adam?" I would suggest. "The same applies regarding your nurses, lab technologists, and all the hospital personnel. It's the built-in home town advantage". If, by this time, they still gave no ground at all, I would suggest to them that maybe we didn't need a hospital at all, that we could just refer anyone needing hospitalization to a physician in Dalsville. Invariably this would meet with a protest such as, "Oh, no! We need a hospital locally for an emergency room in the middle of the night and for people unable to get back and forth from Dalsville," they would protest. By engaging this type patient in this manner, I was able to make progress toward educating them to the fact that the medical system works best for patients and doctors when the patients enter it at the appropriate level for their needs. Over the course of many such discussions I was surprised to learn that the public almost never gave thought to objective issues such as equality of services, value, convenience, efficiency, or the hometown advantage. Instead, they tended to select hospitals on the theory that the best one for them was the hospital capable of handling the most complex and exotic medical need imaginable. Their erroneous concepts smacked of the media's fascination with tertiary care hospitals and exotic illnesses as nightly fodder for TV medical drama series then, as now. If these folks utilized the same mode of thinking for buying a new family truck for hauling their dogs and their garden supplies, they would wind up paying $10,000 extra for a V-8 truck that was too fast for the teenager's safety and too large to get into their own garage.

Probably because of my crusading for the use of the local hospital for things it did as well or better than Dalsville hospitals, I was eventually approached by the medical record department of our hospital regarding any interest I might have in helping the hospital regain its accreditation from the *Joint Commission on Accreditation of Healthcare Organizations* (today called The Joint Commission), that it had ceased even striving for about seven years previously. I was only vaguely aware of that organization and its function in those days, and I naively agreed to help. Retrospectively, I should have paid more heed and respect of the views of my senior medical colleagues on the topic of this organization. Although each of them provided a slightly different example in describing their past experiences with The Joint Commission, the expletives they used in referring to The Joint Commission were always the same, as was the common theme. In essence, it was that regardless what you did at The Joint Commission's request, it was never enough on their next inspection. All my senior colleagues had stories where the next inspection by The Joint Commission brought entirely new recommendations, sometimes even seeming diametrically opposed to the standards utilized on their previous inspection. Eventually The Joint Commission's continuously changing demands, coupled with the doctors' increasingly large time blocks required for servicing its requests, had led the experienced medical staff to conclude that it simply wasn't worth it. Nevertheless, naiveté and youthful optimism led me to jump into it with medical records, and eventually we invited The Joint Commission for an inspection. The day was a total disaster and at their exit interview they made their recommendations for corrections. With patronizing smiles they ridiculed our hastily assembled policies and documentation, and I felt like a young child caught trying to put something over on his parents. In essence, this is what they wanted us to feel, and they made it quite clear that only because they were glad to see some renewed interest on the part of the medical staff due to "youth" (they should have used naivete here), were they willing to give us a one-year-conditional-accreditation with a follow-up visit scheduled in one year's time. That was enough though, enough to make headlines in the local paper and enough Joint Commission experience for me to realize the wisdom of my older colleagues in choosing to save their energy and time for patient care. Retrospectively, I'm glad that I did not know then how the future would find The Joint Commission entrenched in the minds of hospital administrators as omnipotent for expediting successful CMS (The Centers for Medicare Medicaid Services) surveys and thus continued access to Medicare dollars. Had I have known, my guilt today at my efforts expended for achieving reaccredidation for that hospital would have been more than I could bear.

The local news coverage of our renewed Joint Commission accreditation propelled me to a physician-member seat on the hospital board, and with it came some unique insight as to what lay ahead in the future for medicine as a profession. At the same time as my appointment to the board, a young civil engineer and a young vice president from a local bank also took seats. This was significant since other than one progressive middle-aged manager of the local Ben Franklin store, the hospital board prior to this had for years consisted of retired farmers with an average age of seventy-five years. Within three months, the younger contingency of the hospital board came to the conclusion that we did not possess the technical expertise for implementing the progressive changes we envisioned for our hospital, nor the legal skills necessary for addressing the burgeoning regulatory demands that Medicare money and our newly acquired Joint Commission accreditation brought with them. At that time in the mid-70's, hospital management as a business, was just off to a running–start, and like many other local hospital boards, we saw them as an easy solution for expediting the transformation of our old hospital into an up–to-date physical facility and most importantly, as an opportunity for turning over the increasing burden of documentation required at Medicare's and the Joint Commission's behest, to someone better equipped for dealing with it. In hindsight, had I realized then that I was ordering the first of many debilitating blows to be delivered by the concept of hospital management corporations, which ultimately would lead to the demise of medicine as a profession, I would never have cast my vote as I did, for Hospital Corporation of America, one of the first and soon to be one of the largest hospital management businesses in the country, to manage our hospital; had my vote been different, however, I would never have had the numerous enlightening experiences or gained at such an early age, insight as to where the practice of medicine was headed at such shocking speed.

HCA representatives had been visible throughout the community for the last three days. Reports and town gossips had them interviewing the locals in the McDonalds and also meeting with the CEO's of all our local industries. Additionally, the word was that they had accessed the business office data from nearby hospitals, particularly the ones in Dalsville. They were interested specifically just how many local residents were utilizing surrounding hospitals in preference to the one in Levelton. Hearing of such thoroughness pleased the medical staff, as we had expressed to them at our first meeting that we felt we had an unjustified local public image problem. After three days of surveying and investigating, they were finally ready to report to the medical staff their findings and their proposal for addressing the problem. Our medical staff meeting was moved to a larger room for that evening, and although the room was unusually quiet, there was a distinct essence of excitement in the silence. All the medical staff was present, and seated in a cluster around one end of the table were four young men from HCA. I do mean young, as the average age must have been only thirty, if that. They were attired in three piece business suits, all of which were tailored, and even the nails of these young business mogul aspirants appeared professionally manicured. One of them rose, introduced himself and his colleagues and with the confidence of a Wall Street bank president, began a presentation of the company's origins, hospitals managed to date, company philosophy, and goals. Having expertly and eloquently dispensed with that task, he moved quickly to the topic we had broached with them at our inaugural meeting. Reminding us that he promised he would have an answer for us tonight as to whether we indeed had a public image problem, he began by explaining the means and methods and hours of work that had gone into the findings we were about to learn of. More than impressed at the thoroughness and ingeniousness demonstrated, we sat with abated breath as he announced,

"You gentlemen indicated to us that you felt you might have a public image problem with the hospital, and indeed, you do! I don't think any of you realize, however, the magnitude of it." Pausing for effect, he resumed by citing the numbers of people personally interviewed and then suggested that we listen to some of the direct quotes they had encountered over the three days of interviewing. "I would like to read for you some of the responses we received when asking about your hospital. Remember now, the following comments are verbatim as we heard them," he prepared us:

"I wouldn't use that hospital if it were the only one in three states."

"I told my wife if she ever thinks I'm having a heart attack, I want her to put me in the car and drive as fast as she can for Dalsville."

"I wouldn't put my dogs in that hospital."

"That hospital killed my grandparents."

The litany of disparaging comments continued through twenty-five in all when, assuring us he could continue with an hour's more if we wished, he wound his presentation down.

"Yes, gentlemen, I'd say you have a definite public image problem." This provoked a hollow titter from us all. We had suspected it was bad, but the venomous nature of these comments really hit us hard. The young man having now our full attentions, delivered his coupe' de grace, "Now", he resumed, "I'm going to go back over that list of twenty-five responses and this time give you the source of the comment, not by name but by description." The silence was deafening as he began with the first one of those previously read comments. "Number 1 came from a nurse in your hospital. Number 2 came from a person working in the pharmacy of your hospital. Number 3 was from an employee in the hospital's kitchen. Number 4 was from an employee in medical records." he announced. And so it went, with all twenty-five of the disparaging comments having originated from within our own hospital. To paraphrase a famous military quote, "We had met our enemy and it was us." The hospital board was as impressed as the medical staff, and the contract with HCA was signed later that same week.

Looking back at those boom times for the growth of hospital management as big business, hospitals and doctors alike were soft and easy sales for these quick-minded business types. Doctors wishing to have as little to do as possible with record keeping, business matters, and management issues, were approached by these organizations with the irresistible explanation, "We know that you doctors want to be free to do what you enjoy and do best, practice medicine. We can do that for you. We can handle the headaches of the day-in and day-out management of the hospital's problems, satisfy the Joint Commission and the Centers for Medicare & Medicaid Services, and leave your time free to see patients." Their sales pitches to the local hospital boards, however, was somewhat different and probably would not have been nearly as effective ten years earlier. Ten years earlier the day-in, day-out management of record keeping and regulatory compliance issues were not felt to be beyond the scope of the local board and local employees. It was primarily the advent of Medicare with its accompanying monstrous demands for record keeping and the legions of other bureaucratic snafus they brought with them, that led to a sense of hopelessness of ever being able to come to grips with it all by a local hospital board meeting once a month. Therefore, the nascent medical management corporation's pitch to the hospital board was simply,

"Look, we have all the resources behind us for dealing with the state and local government demands, the expertise in every field, and after all, you don't have to do what we tell you. We simply use our resources to give you the best information and hope your boards will follow our recommendations. If it doesn't work out for either party, we can part company and dissolve our contract."

After the first few months with HCA at the helm, what you thought of their service depended upon what hat you wore. Even then, as in my own case, mixed feelings could exist under the same hat. They were incredibly good at what they did, and their resources were tremendous. Their volume buying contracts alone saved our small hospital over a hundred thousand dollars the first year. If you were considering an emergency room expansion, they could have a contractor present at the next board meeting who did nothing but design and build emergency department infrastructure throughout the USA. In theory all decisions had to be made and voted on by the local hospital board. However, they were so persuasive when making their recommendations that invariably the board voted in favor of their proposals. From my perspective the good they accomplished was the new clinical services they instituted such as full time ER physician coverage, physical therapy, and respiratory therapy. The negative side was that like the Joint Commission of the future, the complexity and size of their organization mandated a precisely accurate paper trail in order for their regional directors and other big-wigs of the organization to drop in, spend a day, and easily find everything at their fingertips. For all the hospital staff, this meant less time spent with patients and more time in committees generating a paper trail for HCA. Another negative was their totally business orientated manner of staffing the hospital. The buzz word was units. Everything, it seemed, was referred to and planned for in units. Unfortunately this was not simply a matter of becoming used to new terminology. It seemed to many of the staff that their partiality toward unit, rather than calling a nurse a nurse or a housekeeper a housekeeper, was based on the fact that neither a nurse nor a housekeeper could be cut into fractions, while units could. Thus it seemed that by cross coverage and other slight-of-hand, they could always arrive at a number of units necessary for a job or staffing a given area of the hospital, that was much lower than we would have anticipated being necessary for assuring the job be properly done. Making matters worse, all too frequently their total unit number included a fraction. If their estimate was that 10 and one-half RN's were needed on the 3-11 shift, then we got 10 RN's, since you could not cut an RN in half. The end result was simply that fewer staff did more work and some work didn't get done as often or as well. Unfortunately, as increasingly the case today, the cuts usually occurred in areas of direct patient care and not in areas responsible for maintaining the paper trail (administrative areas).

If the schedule showed 10 RN's on the 3-11 shift, at least four of these would be tied up by supervisory and other functions associated with generation of the paper trail necessary for the firm to keep abreast of operations from a distance. Had I been more of a visionary in those days, I would have recognized from those early experiences with a hospital being managed by business interests rather than professional and altruistic ones, the tremendous malignant potential of hospital management firms for replacing medicine the profession with medicine the business and for impacting the quality of care negatively.

Over time, my primary function on the hospital board became that of physician recruiter, and my experiences in this arena along with the numbers and types of ER physicians HCA was providing us, led me to some disturbing conclusions regarding job security for physicians of the future. I was more than a little surprised to find that already in the late '70s it was a buyer's market where recruitment was concerned, particularly for the specialties. If we indicated we were interested in a general surgeon, over the next month we could expect a number of our evenings to be tied up with wining, dining, and touring any number of candidates. In getting to know these candidates, another disturbing trend came to light. For the most part, these were not young, newly graduated MDs coming from U.S. schools. Most were over 35 years of age and usually of foreign origin. The few American candidates we encountered were usually even older, often over 50. The common denominator for them all was that they were coming from large urban areas.

One fall weekend in 1977, I had stopped in the hospital cafeteria for the obligatory coffee necessary for getting through my rounds in timely fashion when I noticed a strange face, the supporting body of which was dressed in surgical scrubs. He had his own requisite coffee in one of his hands, and the other was occupied in leafing through the local newspaper. He had thinning, gray hair, wrinkled brow, and an air of self-assurance about him. I guessed him to be in his early sixties, but I later learned he was only fifty-five. He introduced himself as Dr. James and confirmed he was working a weekend shift as our emergency room physician. He was very well spoken and indicated he was from St. Louis and still had a private practice there. Curious as to why an established doc of his years was hundreds of miles from home working the weekend shift in an emergency room, I began by inquiring what type of practice he maintained in St. Louis. Fortunately for us both, I had already swallowed my first gulp of coffee when he answered,

"I'm a board certified hematologist" (a specialist in disorders of the blood). Suspecting I knew already my next question's answer I quipped,

"Let me guess, you get a bit bored with the same type patients every day so you do occasional ER work to break the monotony and to keep your hands in medicine in general." Grinning now he pulled his wallet out and was digging for something. I changed my mind and was now expecting him to show me a picture of his late wife and to tell me he was trying to occupy his free time while coping with his loss. He located what he was after and handed me a picture. As I tilted the photo toward better light, I heard him add,

"I like to paint her about every eight years and at her age there are a lot of other upkeep expenditures as well." I was staring at a huge, three-storied, old Victorian home. It was gleaming-white and a magnificent piece of architectural history. Dr. James then continued his explanation as to why he did emergency room work part-time. In a nutshell, it was to maintain a lifestyle he and his wife had grown accustomed to when his practice income was much greater. He went on to explain that when he first had begun his practice of hematology, he was one of only four hematologists in the entire St. Louis area. Today, there were four hematologists on staff at his hospital alone. Not only was that a problem, but there were so many general internists on staff at his hospital that the usual opportunity for bringing new patients into one's practice through an emergency room call schedule was no longer working, since he was only on call to the emergency room about once a month. The frightening simple truth was that he was doing the Levelton ER because he wanted to have his home painted next year! Already as early as the late '70s, over-doctoring within his desirable geographic area had resulted in less income, making it necessary for him to migrate away from his urban surroundings and become involved in an area of practice that due to his sub specialized training, he was less familiar with. The significance of this latter point was brought home to me through one of my own patients seen by Dr. James that weekend in the emergency room. The patient had gone to the ER for a corneal abrasion and was examined by Dr. James and given a prescription for ophthalmic drops containing a steroid, something ophthalmologists and good primary care physicians usually do not do. Not because he was an incompetent physician but because he had extended his practice to a broader scope that he no longer kept abreast of in his reading and training. Most significant of all, is the fact that he did this because increased numbers of physicians had led to diminished workload and income in his sub-specialty practice. Dr. James's solution to over-doctoring in his sub-specialty held profound implications for the future where healthcare providers of all types would abound.

Although Dr. James was the first physician to introduce me to this pattern of physician migration resulting from increasing competition, the pattern became very familiar to me while I was in charge of the recruitment there. As disconcerting as it was, I recognized that the much discussed and ballyhooed doctor shortage, which made the future seem so secure for us, was more correctly only a mal-distribution. Basically the migration pattern went like this. The newly graduated MDs from U.S. medical schools and the newly licensed foreign MDs, located in urban areas where lifestyles were more upscale. Eventually these more desirable lifestyle locations reached a physician saturation point, and the overflow would spill outward to the next most desirable location and so on. With time, the most desirable of these locations would become so saturated as to actually introduce competition for patients among certain specialties, forcing some doctors to relocate to areas with more open markets. For a rural physician recruiter, this basic migration pattern held a pertinent and important corollary. This corollary could better be expressed by the question, "If this doc wasn't able to survive the competition of the saturated urban market, why wasn't he?" Certainly back then as now, there were a few physicians who preferred the less stressful, rural life. However, most of these bucolic individuals located in rural areas fresh out of their training. On the occasion when a U.S.-trained specialist of more advanced age would give us a look over, it paid to remember the competitive survivor question, although it usually turned out that they were simply looking for lighter workloads, having already achieved their economic security. Unfortunately for me as a recruiter, a doctor seeking lighter workloads usually translated into part-time office hours, not exactly a recruiter's dream, but one that in certain instances, was workable for us. Through later years and to date, I have found this basic mal-distribution and overflow pattern to be valid, and with the physician supply of today the pattern is more reliable than ever.

Throughout this managed care transition period of our small community hospital, our practice remained extremely busy and although demanding, still fun. Looking back from today's big brother-monitored and doctor unfriendly atmosphere, it's the memory of those days, when medicine still was a *profession,* which sustains me. In those times we actually felt hospitals, nurses, and medical records personnel were our allies, rather than self-serving paper trail and policy-generators.

A lot of work got done in our offices back then, involving a lot of memorable patients. I could never forget Mrs. Johnson, an 86 year old little lady whose mind and most of her body continued to serve her well. Her only physical cross to bear was a completely prolapsed uterus, unless held in place by pessary and tightly applied under apparel. The condition had existed for years and despite encouragement from us all, she would not consider having a hysterectomy. It simply was not an option from her perspective. On the first occasion that I examined her I nearly flew off my exam stool. Lifting the drape and switching on the exam light I was dazzled by what looked like a snow-white squash gleaming and reflecting the light back at me. It turned out that the meticulously hygienic Mrs. Johnson always powdered her uterus before visiting her doctor.

Another memorable patient, also female but of much younger vintage, provided us with a momentary shock and a good deal of laughter after the fact. Susan was in for her annual pap smear, pelvic exam, and a refill on her birth control pills. When I positioned myself at the foot of the exam table and directed the light onto the anatomy presented to me, there perfectly orientated, as if a seal placed with the intention of preventing opening of her labia, was a U.S. postage stamp. I managed to gently remove it while inserting the vaginal speculum without her ever suspecting that it was there. Afterwards when my nurse and I had finished our laughter, it took very little speculation to figure out how this had occurred. It was my nurse who figured it out. According to her it was very common for ladies, as a last minute ritual before coming down the hall for their pelvic exam, to dart into the nearby ladies restroom for a last emptying of their bladder. We checked, and sure enough, on that particular day we had managed to let the bathroom tissue run out. It took little imagination to picture this harried lady frantically riffling through her purse for a tissue and in the process, inadvertently preparing her vulva for mailing.

We didn't just enjoy our patients, we learned from them as well. It was one of our patients destined to be always referred to in the future as the "cricket man" who taught us the importance of communication and of never talking-down to a patient. The "cricket man" was a man in his mid-seventies whose cast-off clothes, poor hygiene, and stringy, gray hair would have made him an ideal model for a street person today, except for the fact that he lived in his deceased parents' modest, ramshackle bungalow on one of the lesser side streets of Levelton. By any other account, however, it was only this property ownership that excluded him from the street people ranks. His bungalow was heated with coal, and it was often debated whether or not the building had access to electricity. Upon entering the exam room I quickly sized him up and thought to myself, this man may very well have a serious medical condition, as it was uncharacteristic for this type of profile to present himself voluntarily to a physician's office. "Well sir, what could I help you with today?" The old man raised his eyes to mine and succinctly muttered,

"I got crickets in my ears."

"Do they sound like a buzzing or is it more like a ringing?" I asked.

"It don't [sic] sound like nothing. I can't hear too good [sic] out of either one of them. They itch, and one of them is beginning to hurt a little." Grabbing the otoscope from the wall I confidently directed the beam of light into the patient's right ear in order to better see the ear wax buildup that I was confident was causing the ringing or chirping in his ears and the itching and pain. What I encountered, however, was not cerumen but a mushy white material occluding the auditory canal.

"Have you tried to use cotton to clean your ears out?" I asked him.

"Nah", he snapped.

"Well, let me rinse some of this away and see what it is then." Having flushed a small bit of the white material into the irrigation pan, I held it forth for him to see and suggested, "All this white stuff, it's not cotton you've had in there?"

"Oh no, that's old bread," he replied matter-of-factly.

"Bread, why would you be putting bread into your ears?" Again succinctly and matter-of-factly, he answered,

"To feed the crickets." Switching quickly now to my learned physician self I suggested to him,

"It may sound like chirping, ringing, or buzzing of crickets, but surely there are no crickets in your ears! His reply was calm and to the point,

"You'd better go on ahead and clean some more of that bread out and see. They're probably dead, but they're in there." After several more laborious minutes of irrigation, interrupted only for brief sojourns outside the exam room door to get my breath and permit my nurse's nausea to be controlled, we eventually evacuated what appeared to be various body parts of a cricket. Further questioning revealed that initially, a friendly cricket developed as a pet, began to spend the night in the external canal of his ear. Apparently the old man felt the physical irritation a small price to pay for having some company, and he permitted this to go on. Eventually he began placing small bits of bread crumbs into the ear canals in order for the cricket to have some nourishment. There was no mistaking it. That man had "crickets in is ears"!

For a while back then, we kept a list of the more exotic cases that we encountered. These were the extremely unusual cases that year's of extensive health science training in medical school covered but were of such small likelihood of ever being encountered, that you never ever dreamed you'd see any of them! During a three year span we encountered two pituitary tumors, one idiopathic ascending sclerosing chlolangitis, a complete vaginal septum, and an acute case of malaria, to name several I still remember from the list. There were others as well, but unfortunately, I misplaced the list over the years. In looking back, I don't wonder about the cases on that lost rare diagnosis list that I can't remember, but rather about any rare diagnoses I might have missed during that same period of time and never knew it. For a primary care physician, the mission and purpose of his broader scope of training is to triage large numbers of patients, effectively diagnosing and treating the majority of them, while screening out and referring to appropriate non-primary care specialties, the very few exotic and complicated conditions requiring specific non–primary care treatments. No other area of medicine requires a physician to more frequently call upon the extensive medical science knowledge base acquired through his years of schooling, than primary care. Today, thanks to special interest groups, we are permitting individuals to attempt this high-risk job of triaging, armed in many instances, probably with less formal basic medical science education than their high school science teachers possessed.

If six elderly patients with ankle swelling are seen during the course of the day, the challenge for a primary doc is to know which one or two of the six might be due to congestive heart failure. It should also be the primary care physician's goal to select the one or two cases due to congestive heart failure out of the six in an economical and efficient manner for the patient. If a Doppler echocardiogram were immediately ordered for all six patients, the two with heart failure would be found, but the needless cost in using an echocardiogram as a screening process for the others having chronic venous insufficiency or other non-cardiac causes of their ankles swelling, would make this method non-justifiable, certainly in the opinion of any patient paying out of pocket for such a service. In order to consistently be successful at selecting out the two cases of ankle swelling due to congestive heart failure, the primary physician must have a medical science knowledge base of sufficient scope and depth for considering all possible physiologic mechanisms which might result in fluid excess in the lower extremities. Once suspected, those two could have a confirmatory echocardiogram for discerning the degree of the problem. Venous insufficiency, congestive heart failure, renal failure, stasis edema, cirrhosis and hypoalbuminemia all can cause ankle swelling. Only by seeing and interviewing the patient, while matching each of these potential causes to the history and physical findings of that particular patient in front of him/her, can a primary doctor efficiently and economically arrive at the proper diagnosis. Indeed, this is what you pay a doctor's fee for, and what only a properly trained MD can provide. After all, Wal-Mart certainly has the resources for purchasing and maintaining MRI units in each of their stores. In fact they probably could offer an MRI scan of the head at a lesser cost than is being currently charged by hospitals.

However, there is an inverse relationship between the depth of the medical education of the individual doing the diagnosing and the number of tests ordered for confirming a diagnosis. A third year medical student just beginning his clinical years of medical school and seeing a patient complaining of right lower quadrant abdominal pain, will utilize many more lab studies and x-rays for assuring himself that he is not missing the proper diagnosis. A seasoned surgeon in private practice would make the diagnosis of appendicitis by history, examination, and perhaps a CBC and urinalysis. This is an exceedingly important concept with regard to cost of medical care issues at the millennium, where a number of physician extenders and paramedical personnel are being utilized as diagnosticians. The needless dollar contribution to the overall cost of medical care for routine matters by these individuals is tremendous, due to their indiscriminate and over-use of ancillary testing. It's frightening and paradoxical that the trend at the millennium, fueled by political, special, and entrepreneurial interests is toward placing minimally educated and minimally experienced persons in the critical role of diagnostician. Should this trend continue unchecked, it will likely add multimillions of dollars of unnecessary testing to the cost of health care annually over the next decade. This trend of utilizing physician extenders as first-contact primary care providers is growing, despite the fact that most non primary-care specialty *MDs* today, are very uncomfortable when they find themselves facing a general medical problem outside their focused areas of practice, regardless how outwardly minor the problem might seem. That's because having undergone the rigorous in depth years of education required for an MD degree, they have an appreciation for the degree of thought that must be given to the diagnostic process, lest the most simple problem turn out to be not what it seemed, but instead, a potentially life threatening masquerader.

Our practice in Levelton in the '70s was truly a primary care one. We saw 45–55 patients every day, sorting them out by history and physical into those for treatment and those in need of referral. Often on the phone with sub-specialists in Dalsville, it seemed sometimes like we referred everything. Retrospectively, it was a very small percentage of our total patient numbers that got referred. Percentage wise, it was comparable to academic and government figures often quoted in those days. Back then it was often quoted that ninety percent of a person's total healthcare needs, from cradle to grave, could be adequately met by a traditionally educated primary care physician. Even with the technological advances of today that have become a standard in today's care such as endoscopies, colonoscopies, colposcopies, etc., this figure would still probably remain at eighty percent or better.

For primary care physicians, the history and physical exam remains their most important tool. Of the two, the history is by far the more important. If the patient can communicate, and if you listen long enough, eventually he will tell you what is wrong with him, provided you have the proper medical education and a pathology background extensive enough for recognizing the signs and symptoms caused by each disease process. An oft used teaching axiom in medical school goes, "If you don't know the patient's diagnosis by the time you finish interviewing him, you probably won't get it from the physical exam." The history is extraordinarily important in the diagnostic process, and there is nothing the equivalent of hearing it in the patient's own words. Subtle voice inflections, facial expressions, gestures, and a host of other para verbal communications make this so. This is another area that has suffered with the '90s trend of applying business applications to the practice of medicine.

These days, in an effort to utilize the physician's time in an economically more efficient manner for the firm, i.e., to cut the actual time a doctor spends with the patient, many managed care clinics have gone to a pre-filled out medical history. This is one or two pages of multiple choice questions designed for placing a check mark to indicate the presence of a sign or symptom. When first beginning to be utilized, such patient generated history forms were at least administered in the presence of a nurse who could help the patient interpret the questions, thereby assuring valid information was being received. With further efforts at increasing efficiency, many times now patients are mailed these forms to be filled out without assistance, in advance of their first office visit. Assuming the patient really takes time and effort to comply and assuming he/she is literate and understands the forms, a minimally helpful baseline of information might be obtained in this fashion. However, such forms don't come close to providing the information accessible by an experienced interviewer in a face-to-face meeting.

Reading over one of the pre-filled histories one might find the following, "Do you have any stomach problems?" Yes or No.

"Do you ever have stomach pain?" Yes or No.

"On a scale of 1 to 5, with 5 being most severe of all, how bad is that pain?"

"Is your pain worse at certain times of the day?" Yes or No.

In a patient with an acute peptic ulcer attack, such a pre-filled history doesn't come close to impacting the doctor's diagnosis and subsequent treatment as would a patient sitting in front of him and telling him,

"Doc, I sure hope you can do something about this stomach pain I've been having this last month or so. It began a couple weeks after our layoff at work. We'd been to the bank that day to take out a temporary loan to get my daughter registered in college, and that night it woke me up in the middle of the night. I tell you, it was like a lump of burning coal right in the pit of my stomach. Sometimes it felt like it was boring right through to my back. I tried a little ice water, but it didn't seem to help much so I took a couple tablespoons of my wife's Mylanta and it eased up some, enough for me to sleep at least. Since then it's off and on through the day but always worse late at night. I switched to Tums during the day because they are easier to carry with me, and they help but I have to keep taking them all the time." Frequently the history taking is just that easy. Just give the patient a chance, and they will tell you the diagnosis. Also, if the incredibly effective acid reducing drugs of today were available back then, doctors would have treated that patient for the gastritis or ulcer he/she obviously had, relying on absence of symptoms at follow-up visit for assuring them he was well, and that they had been correct in their diagnosis. Today, the fear of misperceptions by the omni-present, omni-documenting paper trail representatives would likely assure this patient's undergoing an EGD (scope inserted into the stomach) before he could expect any treatment of this problem. If a patient today with such a clear cut history for peptic ulcer disease were to encounter a physician whose background and philosophy leaned towards treatment and follow-up as an effective, practical, and economical alternative; that physician, in the over-doctored competitive atmosphere of today, could easily rationalize performing an endoscopy anyway, in the interest of paper trail requirements or defensive medicine, when increased revenue from the procedure was really the deciding factor.

In certain instances, experience and the medical education have to be called upon in knowing what questions to ask and how to ask them. One memorable patient, whose history unexpectedly provided the correct diagnosis, was referred to me by Dr. Arnold, one of the two young physicians who practiced about fifteen miles from Levelton and occasionally used our hospital. I received a call from Dr. Arnold in the middle of office hours one day for asking me if I would workup a nice little old lady for her GI complaints. Knowing him to be a very competent clinician, I fell victim to my ego at his asking and quickly agreed before hearing the rest of the story. He then added that the patient had been having her complaints for about two years, and that about one year ago, at the request of a daughter living in the St. Louis area, had been evaluated at a medical center in St. Louis without anything being found. Beginning now to chuckle a bit, he admitted that he had seen the patient only a couple of times in his office with her initial complaint before the patient's daughter contacted him and asked him to refer her mother to the large tertiary care hospital in St. Louis.

"Anyway," he continued, "She is back in here today with the same old complaints, so I guess they didn't help her any more in St. Louis than I did," he jibed. "I told her today that I knew a really nice, sharp, young doctor in Levelton who might be able to get to the bottom of it for her," he chortled. And then his coup de grace, "Oh yeah, I'm going to make it easy for you. She has copied her complete chart from her St. Louis evaluation and will bring it in with her. I hope you have a table high enough to fit it under so it doesn't obstruct the traffic flow pattern in your office," he cackled.

Knowing now that I had been had, I dreaded the next day's hospital rounds when I would meet this chronic GI complainer for the first time. Taking a long deep breath and gently pushing open the door of her private room, I introduced myself as I entered. After 15 minutes discussion with her, I left the room in possession of one major attitude adjustment. Instead of the depressed, self-focused and polysomatic, chronic complainer I had expected, this patient was a pleasant, well-grounded, middle-aged lady with only one complaint. Hopeful now and encouraged by the patient herself, I was jerked back to reality when she described her "attacks". Her description of the symptoms were classic for chronic gall bladder disease, but having undergone a complete evaluation at the center in St Louis, she obviously couldn't have anything so simple, it looked like a tough, if not impossible diagnosis. I sat down in a corner of the nursing station and began pouring over the test results from her hospitalization in St. Louis. It was an extensive workup for those days and included an upper GI barium x-ray study with small bowel follow through, a barium enema x-ray of the colon, liver enzymes, an EGD (scoping of her esophagus and stomach) and stool studies. There was only one study that I did not see evidence of in her record, and it was the very study that any physician would normally have ordered first off, given her classic description of gall bladder stones. Try as I might, I could not find any evidence of an oral cholecystogram having been done. This was before ultrasound was available and the gold standard test for gall stones then was an oral cholecystogram. A phone call to Dr. Arnold and to the St. Louis hospital confirmed that no routine study of the gall bladder had ever been performed. The original referral to St. Louis had been made spur of the moment by telephone when the patient's daughter had suddenly pressured Dr. Arnold to arrange it. In relaying to the St. Louis physician Mrs. Johnson's symptoms, Dr. Arnold had apparently used the term classic "gall bladder picture" or some similar phrase The St. Louis physician, being in a tertiary level hospital and not accustomed to getting out of state referrals for common conditions, apparently interpreted the hurried and brief conversation with Dr.Arnold as meaning that although the symptoms were classic, gall bladder disease was not the cause and <u>assumed</u> the referral was for a more complex diagnostic evaluation. In other words, the St. Louis physician heard what he expected to hear. The end result was that the oral cholecystogram we ordered at the Levelton hospital for her revealed stones. Ms. Johnson was ecstatic, and if memory serves me, the patient's daughter was grateful too but still insisted on a referral to a Dalsville surgeon rather than permitting our surgeon to remove her mother's gall bladder while she was there in the Levelton hospital!!

Having said how important the history is for the diagnosis, rare exceptions exist where a look is worth a thousand words of history. One such case for me during those years ended up being one of the most memorable of my career to date, although my role in it was very minor. We had a very frequent patient by the name of Julie. She exemplified what at that time was unkindly referred to in private among physicians, as a "crock". This was slangy vernacular for a chronic and repetitive complainer whose complaints changed more frequently than the weather. Typically such types will begin seeing a particular physician and for a year or so they will be as familiar to his office staff as the office decor. Invariably, nothing can ever be found to explain their complaints, and inevitably, periodic and wasteful sub-specialty referrals wind up being made for the purpose of reassuring the patient. From our perspective referrals of these types of patients were made similarly to the way the non-medical public of today arranges their self-referrals when left to their own devices, i.e., via the anatomic specialty method. If the patient's chronic recurring complaint is a headache, then a neurologist gets the call. If the persistent complaint is constipation and abdominal symptoms, the gastroenterologist gets the call and so forth. Not Julie, but I can recall similar type patients in their frustration and anxiety requesting of me, "Well doc, if you've checked everything you know of, could you refer me to a specialist," with me sitting there frantically searching for a diplomatic reply, since their symptoms were so vague and bizarre as to not even fall into any particular anatomical area! Then as today, non–primary care specialists prefer sending patients with such vague and unclear complaints to primary care physicians. The problem today is it is increasingly hard to find a good primary care physician for sending such patients to, and, sadly for the consumer, there are many forces now at play that will undoubtedly assure that good primary care physicians are facing the fate of the passenger pigeon in the very near future.

In Julie's case, after seeing her as a patient for nearly two years, we had wasted the time of an orthopedist, a neurologist, and an ENT physician. This was as frustrating for me as it was for her, and also sad since she was an attractive 28 year-old that should have been enjoying life, friends, and family. I say family, but my only knowledge concerning her family was that she was married and had one child, a daughter. I had never seen or met either the daughter or Julie's husband. One afternoon as I was passing by the waiting room area I caught a glimpse of Julie's face among the others and briefly wondered which organ system would be the focus for her irrational concern today. A few minutes later I opened an exam room door to see her again but not where I'd expected. She was seated in the chair for visitors, and on the side of the exam table was perched an angelic five year old miniature of herself. Immediately, mom introduced her to me as Tonya, her daughter, and after my greeting, began her story of why Tonya was there. Tonya sat with legs dangling, every now and then straightening one of them out to examine her shoe as her mother related for me how Tonya had just not seemed herself recently. For about a month she had seemed tired all the time, preferring to lie on the couch in front of the TV and often nodding off to sleep after only a few minutes. According to her mother this was entirely unlike her, as usually she was impossible to keep up with. Interrupting here, I asked if there had been any specific complaints or signs during these past months.

"No, she just looks tired and pale," Julie offered.

" Well, who is Tonya's usual doctor?" I asked. Julie scooted a little uncomfortably in her seat before answering,

"Well, I have always taken her to a pediatrician over in Bateyville. This today is the first time she has ever seen another doctor. Given the ideal opening I seized it.

"Why did you bring her to me today then?"

"Well to be honest with you, for a second opinion. I've taken her to Dr. Hart twice in the last two weeks, and he keeps saying there's nothing wrong. Well, the last time he did say he thought she might have a virus." Hoping to myself that Julie was not about to transfer all of her obsessive health angst from herself to this cute little child, I picked up a tongue blade and asked,

"Honey, can you open up and let me have a look?" After a shy smile and a hurry-up last second swallow, she opened wide, and as my light struck her gums my heart sank. Scattered over their surfaces and also over the mucosa of her cheeks, were clusters of pinpoint dark dots. The medical terminology for these is petechiae, and they represent pinpoint hemorrhages due to very low or dysfunctional platelet counts of the blood. None of the possible causes of low platelets was good, and one cause in particular was extremely bad, leukemia! "How long ago was it that you were last seen by Dr. Hart?" I asked.

"Oh, nearly two weeks ago, I think," Julie nervously answered.

"The doctor examined her at that visit?"

"Oh yes, he always does a good exam," she explained.

"Did he order any blood work to see if the viral infection was a bad one?" Not wishing to prematurely scare this obsessive mom, I was trying to lay general groundwork for ordering a complete blood count with a platelet count.

"No, he didn't do any blood work then or the visit before last visit either."

"Well then, let's just do that today for the sake of thoroughness," I suggested. I finished my exam so as not to raise any suspicions until I had the lab data in hand and told Julie to give me a call later that afternoon, and I would tell her how the lab work turned out. I must have covered over my own worry well because she accepted my suggestion without further questions. No more than half an hour later the lab tech was already calling me. This bode ill for Tonya since extremely bad results always were called to the doctor immediately. I was informed that Tonya's white blood cell count was over 70,000 with a very low platelet count and a severe anemia as well. The lab tech said it probably was an acute lymphoblastic leukemia, but of course, he wanted the pathologist to confirm it for him. Sitting down in my office to think about that darling girl's future, I found myself instead thinking of her mother, maybe because I was a father of two young boys myself or maybe because of what I knew of Julie. It was probably both, but without question, I wondered how Julie could possibly deal with a real illness of this magnitude in her daughter. Later that same afternoon Julie, her husband, and I were discussing where Tonya should go for her treatment. There had been the initial reaction of any parent when told of the nature of the illness, but after several minutes both appeared calm and focused. Julie was an entirely different person. She listened attentively and asked reasonable and pertinent questions. Apparently she had family in the Indianapolis area and was familiar with Riley Children's Hospital there and indicated she was leaning in that direction. I told her it was an excellent suggestion but wondered if she had considered St. Jude's in Memphis. In one of those serendipitous moments of life, I had attended a week long Family Practice review course in Memphis the previous month and had seen a presentation by the St. Jude's staff of their cutting edge treatments in childhood leukemias. Their success had impressed me greatly and I had brought home with me their entire folder summarizing the services they offered there. It took less than 10 minutes for telling them of the program and for them to digest the statistics in the brochure. We were all agreed, St. Jude's was the place for Tonya. After calling the St. Jude folks and making the arrangements, I admitted Tonya to our hospital as they had suggested until we received a confirmation call and further instructions the next day. Around noon the following day I said goodbye to them all, and they headed out to Memphis. It was the last time I would see any of them as patients, but communication was never lost entirely due to Tonya's maternal grandmother. Initially she called me after Tonya had been in St. Jude's for about a week. She began that first call as she would many of future calls to come, by tearfully thanking me and singing my praises. Embarrassed, I told her truthfully that I really had only a very small role in it all, explaining to her as I had Tonya's parents that day, that Tonya just happened to be in my office on the day that the petechiae were present for seeing. I told her that had the petechiae been there when Tonya had seen Dr. Hart, he would have ordered the same tests.

"Well maybe so, but it was your idea to send her to St. Jude's and it's been wonderful there," she blubbered, so that I felt embarrassed for several minutes after hanging up the phone. Tonya's grandmother, hearsay in the community, an occasional newspaper article, and a few initial follow-up reports from St. Jude were our only sources over the next year on how well Tonya was responding to therapy. Even after relocating my practice out of state, on Thanksgiving, Christmas, and Tonya's birthday I received thank you calls from Tonya's grandmother for several years to come. On the last of these I told her my son was going to be riding his bicycle in the local St. Jude's pledge bike-a-thon. About a week before the St. Jude event we received a package in the mail containing an autographed national poster with Tonya's picture on it, along with a note explaining that Tonya had been selected the national poster child for St. Jude's!! Also in the package was one of Tonya's St. Jude Hospital ID bracelets that she had worn on one of her many occasions there and a note from her thanking me and suggesting that my son carry the bracelet on his handlebars during the bike-a-thon! After all these years I still get goose bumps when remembering that package from Tonya! For sure my role in Tonya's success was miniscule, but looking back from today's misused, abused and budget-busting medical delivery system, I can rightfully be proud of my role in Tonya's care for another reason. That day when I met Tonya in my office, she received from me exactly what was needed, nothing not needed, at a reasonable price, and with all the convenience in the world.

In today's medical milieu the odds against successfully meeting these needs for a similar patient under such circumstances, would be astronomical. Today, Julie might well be a member of an HMO and therefore be limited to seeing a medical oncologist within that group. Certainly a medical oncologist would be qualified to manage and treat an acute lymphoblastic leukemia in a child, but would an adult oncologist who might see one or two such cases in that age group a year, be the equivalent of being managed in a program such as St. Jude's, which sees hundreds of cases yearly and who's entire staff is dedicated to caring for this particular illness, in this particular age group. In today's competitive arena of profit oriented managed care, even if a mother like Julie knew for a fact that the five year survival rate of patients treated by her HMO for acute leukemia was three per cent less than at St. Jude's, do you think that she could expect to be referred outside of her managed care group because of this fact? Or, if she were private pay and could afford to go anywhere she wished for the treatment of her daughter's leukemia, do you think for a moment if she presented herself to a primary care physician within a large multiple specialty clinic owned and managed by an urban hospital, that the primary care physician would likely discuss with her the top five programs in the nation for treating the disease from a statistical success standpoint? Under such circumstances today it's more than likely, unless she has done her own homework, is financially well enough off to pay her own way, or insists on a referral; she will end up hearing only of the medical oncologists associated with that multi specialty clinic or some other one associated with the management firm running that hospital.

In years past, before any type of managed care, this referral within the group trap was well known as a potential problem when utilizing multi specialty groups for referring sub-specialty cases to. Family physicians then, considered it a vital part of their services to see that their patients did not fall victim to it. The problem resided in the fact that once being seen by a particular specialty within a multispecialty clinic, should the need arise for consultation from a different specialty, the referral automatically was made to a specialist also within that same group practice. Unless the multispecialty clinic was one of impeccable, national reputation, with top quality physicians in each and every one of their various specialty departments, the potential always existed for being referred to another physician within that clinic who might possess only average, or maybe even less than average, credentials and experience. For the few venerable multispecialty clinics of impeccable reputations, recruitment of outstanding physicians for all their specialties was not a problem. Like medical schools in the sixties, whenever a position opened within their ranks, such renowned organizations were deluged with outstanding candidates to choose from. These were institutions such as Cleveland Clinic, Ochsner Clinic and Mayo Clinic, to name a few more familiar to the public in general. For all the other large clinics that were calling themselves multispecialty clinics, this was not necessarily the case, and they often had to settle for mediocre or even less, if they were having trouble filling a particular specialty slot in their group.

For example, a self-referred patient suffering nausea and vomiting might decide they need a good gastroenterologist due to the nature of their symptoms. Wanting the very best, they might investigate thoroughly and learn that Dr. Smith, a gastroenterologist with the Valley View Multispecialty Clinic has impeccable credentials and an outstanding reputation among his peers. An appointment with

Dr. Smith would then be obtained and after seeing them, Dr. Smith might be extremely suspicious that the patient's repetitive vomiting is due to increased intracranial pressure, possibly secondary to a brain tumor. He logically would immediately then make her an appointment for the same day with Dr. Stams, a neurosurgeon also with the Valley View Clinic. Unfortunately, however, the Valley View Clinic's neurosurgical department had been difficult to recruit for, and after many attempts they had finally gotten Dr. Stams. At the time, none of the clinic's credentialing committee was particularly impressed with his educational background, experience, or previous clinic associations, but he is a board certified neurosurgeon, and that position within their clinic has been vacant far too long. End result, one specialty department represented at the Valley View Clinic is perhaps mediocre or less compared with the clinic's other departments. As a result, the vomiting and discriminating patient who had so carefully sought out the best gastroenterologist, may end up being operated on for a brain tumor by a so-so neurosurgeon of limited experience, while two excellent neurosurgeons of national reputation might be located only an hour's drive from the Valley View Clinic.

In past years, it was considered a routine but crucial part of good primary care, to educate one's patients regarding such pitfalls in the medical system. It used to be considered a professional and ethical responsibility to send one's specialty referrals to the best possible physician available for meeting their patients' needs. In other words, the best primary care doctors in past times applied the golden rule in referring patients. If they wouldn't want their own family members seeing a particular doctor, then they didn't refer their patients to that physician either. In those days, the best primary care doctors knew who were the best sub-specialists and vice versa. Each recommended the other when the need arose. Medicine then was still an honorable profession percolating with pride, and to be held in high esteem, by colleagues, highly esteemed, was one of the most rewarding aspects of it. To be thought of as one of the best thinkers was a kind of carry over from the competitive one-up-man-ship of the clinical medical school years and, believe it or not, was more important than financial success to most physicians then. In those days medicine was a proud profession.

Most of my colleagues from the '70s would find the conditions and watchdog atmosphere of today's medical arena intolerable. One professional I met back then, however, would have fit right in with the excessive paranoia of today. Mick Holloway was his name, although among local docs he was referred to as Big Mick. He was a large man, being well over 6'-2", with weight to match. He also was Levelton's oldest living pharmacist. His drug store reflected his senior standing well. The other drug stores in town had years before remodeled their interiors with chrome-edged glass cases, low white metal shelving, and recessed fluorescent lighting. Big Mick's store, however, was a relic of the '40s, having ceiling-high dark wooden shelves, marshmallow-shaped white-globed ceiling lights, thick-glassed display counters edged in dark wood and lighted with exposed bare light bulbs, and a large stoppered five gallon decorative glass flask, standing sentinel duty at each end of the prescription counter. One of the flasks contained red liquid and the other green. The colors were not true, however, due to the layer of dust on them and the absence of any reasonable light in the entire store. As memorable as Big Mick's store was, the man was more so. His business volume was small relative to the other pharmacies in town, so proportionately very few of our patients' prescriptions were filled by Big Mick, but we knew every single one of them by heart.

The first several of Big Mick's calls back to me, I attributed to my being new in town and my handwriting new to him. Since his clients were few, it took nearly a year for encountering enough of them in the office for realizing that he called the office and read back to me every single prescription that I wrote. He was so consistent that if a patient leaving our office drove directly to town and found a parking place, we could predict within a few minutes when our phone would ring. Every callback was the same and went like this.

"Uh, Doctor Daley, this is Mick Holloway down at Holloway Drugs, and I have one of your prescriptions for Mr. Bob Smith. Now you did want…", and then he would read verbatim my prescription back to me. Maybe if you spoke with him only once or twice a year, his voice and tone would not have seemed noteworthy, but hearing him once or twice a week was sufficient for magnifying all its annoying qualities. It was a deep monotone, and each word was groaned out as if he were studying and contemplating it for some hidden meaning. No matter how trivial the content, Big Mick pardoned no prescription from this ordeal. A script for a laxative was treated exactly the same as one for a cardiac drug. Eventually, I was gratified to learn that it wasn't me Big Mick was questioning but himself. He afforded the same annoyance to every other physician in Levelton as well. For whatever reason, none of them had ever refused his call over the years, although all of them joked and made fun of him. Office nurses routinely competed in Big Mick imitations after one of his calls, and a couple of them were quite accomplished at it. Your car might not start, the power might fail, a year might pass without rain, but Big Mick never filled a prescription without calling the doctor to double-check it. About this same time, Dr. Sexton's problems seemed escalating in frequency, and on some days his office would never open for his patients, although it was evident that he and his nurse were there from their cars parked out back of the building. His office nurse was Jan Smith, an LPN who had worked several years at our hospital before leaving under a cloud of rumors concerning missing narcotic doses. She was an obese girl in her late twenties, outgoing and loud in a pleasant sort of way. She always had a greeting for me and frequently would strike up a conversation as if we were old acquaintances. In reality, I never had talked with her outside the hospital environment and even there only for replying to her usual light banter. Early in the afternoon on a particularly busy office day, my nurse approached me with a message that Big Mick was on line two.

"We haven't seen any of his customers today have we?" she asked as I was punching in the blinking line 2 button.

"Dr. Daley here," I said into the phone and waited for Big Mick's routine.

"Dr. Daley, this is Mick Holloway down at Holloway Drugs." Then, an unusual break from his routine by a couple seconds of total silence followed by, "I got to worrying a little that some of those dilaudid tablets (a narcotic analgesic of high abuse potential) I gave Jan this morning had been on the shelf a long time. I guess they could be outdated by now. Would that make any difference?" No question, it was Big Mick's droning contemplative voice, but all I could say was

"What!"

"Those dilaudid tabs you wanted for Jan Smith's mother. Some of them could have been two or three years old that I gave her."

"Mr. Holloway, I didn't write for any dilaudid for Jan Smith. She doesn't even see me as her doctor." Now, there was a really long silence followed by,

"Well, no, she didn't have a written prescription. She had just gotten back from Dalsville where her mother is being treated for metastatic cancer, and she told me she had called you, and that you said come by here and pick up sixty, 2 mg. dilaudid tabs for using one every three to four hours for pain. I could see her mother sitting out in her car," he added hopefully.

"I'm sorry, I really don't know a thing about it," I had to tell him.

"Oh me, what am I going to do now. She told me that her mother was in so much pain that you told her to stop here on the way home and get the dilaudid so she could get her mother on home, and that you would give me a script later." Sensing his panic at the realization of what he had done, I assured him I would take care of it and get back with him. First, I called Jan Smith at her home. I identified myself and without further questioning simply told her that I understood that she had something in her possession that she had obtained by using my name, and I would be over to pick it up. She nervously denied knowing what I was referring to, and I curtly suggested that I believed she did and hung up the phone. Then I called the local police, and a detective came by the office. I filled him in on the story, telling him I was not interested in pressing charges, but that I wanted to retrieve the dilaudid and get it back to Big Mick. The detective, a long time resident of Levelton and veteran of the force, thanked me and assured me he would take care of it. Less than an hour later he was on the phone and telling me he had rung Jan's doorbell, and when she came to the door he said simply,

"Dr. Daley sent me to pick something up that has his name on it." According to the detective she had wheeled around, walked into a small bathroom just off the entry hall and returned with a bottle containing the sixty dilaudid tabs. "They're already back at Holloway Drugs," he said.

Who would ever have thought that Big Mick, of all people, would have been the pharmacist involved in such a fiasco? Apparently on that particular day, his obsessive compulsive personality must have taken several minutes of much needed rest, after years in overdrive.

My position as physician-member of our HCA managed hospital board made me privy to information and trends that came increasingly to worry me with regard to the future of medicine, certainly as a profession and maybe, even as a career. It seemed obvious to me that the unabated influx of foreign physicians from all over the world, the continuous ratcheting upward of class size by US medical schools, and the incessant but inaccurate (at this time) clamors from the media of doctor shortages would eventually lead to a surplus of physicians. A potential physician surplus, partnered with the increasing numbers and clout of hospital management businesses, in my estimation, bode ill for the future of the medical profession, physician job security, cost of care, consumer satisfaction, and probably quality as well. This was

1980. With these concerns on my mind and still missing the geography of my home state, I decided to return there while there was still time enough for establishing a new practice before these ills took their toll in earnest. Since this was back when the non-insured still had access to medical care (thanks to doctors carrying them on their books), I left Levelton with a wealth of personal experience and a small fortune owed me on the books.

CHAPTER 3

The timing and motivation of my relocation were sound, and if only I'd done as well with my choice of places, the plan would have served me well. Instead, I would spend the next eight years in a small Kentucky town which, because of long and bitter feuding over power within its school board and hospital board, prematurely represented a microcosm of all my worst fears of what could result from future physician surpluses and hospital managements with agendas other than their patients' best interests. Seduced by its idyllic geography, I unknowingly turned forward for me the timetable of medicine's professional demise by relocating smack in the middle of this small town's bitter power struggles. The roots of the feud were old and deep, and the central characters were a 70 year-old surgeon and his clinic, versus the current local politically empowered. The primary battleground was the just completed, new community hospital, and the heart of each side's battle plan was physician recruitment. The battle strategies of the two sides were identical and consisted of gaining majority representation on the medical staff of the new hospital in order to make life as miserable as possible for the opposition. In my eight years there I saw numerous physicians come and go (around 20 by memory). Physicians were not the major casualties, however, that was reserved for the hospital, consumers and the taxpayers.

I had noticed a physician recruitment ad in one of my professional journals and subsequently decided to visit the area on my own. I stopped by the hospital, introduced myself, and was told by hospital representatives that indeed they were in desperate need of doctors, as all the town's other physicians were "getting up in years" and would welcome a new face, as they were all just waiting to retire.

"So, the other doctors agree that another physician is needed here?" I asked point blank.

"Oh, yes. The doctors here are all good ones and easy to work with. Well, all but one. There's old Dr. Bailey who's been here for years, but his practice is losing patients every day. He's always difficult to get along with and always stirring something up. To tell the truth, I think he might be getting a little crazy in his years. Says he's going to remodel the old hospital building for his own clinic. Isn't that crazy? You don't need to be concerned with him though, he's winding down and about ready to hang it up." It was after five when I finished driving around the town and talking with the hospital representative, but I decided, that crazy or not, anyone who might be remodeling an entire hospital building did not seem likely to be "hanging it up", regardless of his age.

I located Dr. Bailey's current office directly across the street from the vacated old hospital. It was in an old, small, aluminum-sided residence that had been more commandeered than converted, for use as a doctor's office. Waiting for him in the reception area, I hoped that Dr. Bailey would be more impressive than his offices. He was. Although this aged, bulldog-framed, country surgeon would cause me no end of political, professional, and personal strife over the next eight years, I liked him at that first meeting and left there, many difficult years later, still liking him.

Unfortunately, I could never say the same for the new hospital or particularly, the hospital's administrator. I disliked and distrusted the administrator from our first introduction, and for the years he continued there, he never once gave me cause for upgrading that initial impression. He seemed smarmy, and his lack of integrity hung over the entire hospital like a monstrous black shroud, without a single gap in it. My first clue should have been when he asked me never to mention to anyone around the hospital the fact that the hospital paid for my moving expenses. Many other clues followed quickly, and within a couple of months, it seemed likely to me that he was the new hospital board's hand picked administrator whose primary agenda had little to do with the new hospital's success and everything to do with the planned destruction of Dr. Bailey's practice and clinic. After a couple of years, when we both knew where the other stood, he finally admitted this to me. He had asked me out to lunch in order to persuade me to quit walking the middle ground and telling the truth regarding his and the board's feud with Dr. Bailey. Having failed again as on numerous past occasions, his unctuous style suddenly ceased, and he threatened,

"Dr. Dailey, you don't know these people here like I do, and I can make your practice here or I can run you out!" Totally honest but less confident than I sounded, I leaned into his face and asked,

"Hiam, you are a city person. Everything about you suggests that. I know you don't like this small town lifestyle, so why are you here?" Voice lowered and lips tightening, he hissed,

"I was brought here by certain people in the county to do a job and when that job's done, I'll be out of here."

"Do you have any idea how long that will be?" I asked.

"Just as soon as Dr. Bailey and his clinic fold," he shot back. After two years in the bizarre hospital atmosphere he and his co-conspirators had created, I already knew this, but it was the first time I heard him actually admit to it, and I got a certain satisfaction from it.

"Well, Hiam, I was having some thoughts about leaving this fiasco you've created here, but I just this minute decided to still be here when you've gone, just for enjoying the improvement." I eventually made good on my response to Hiam's threat of that day but at an unreasonable cost to my finances, family stress level, and trust in people.

Within the first six months in my new community and new practice, I had confirmed to my personal satisfaction, the nature of the battle I had unwillingly stepped into. Basically it was one of dominating politics and dominating personalities that typify poor rural towns whose major employment is provided by the school system and the hospital. What were unique in this case, though, were the battle's ubiquity, intensity, and its staging area. There was no group, organization, or individual that did not vehemently support one side or the other and spend part of every single day gossiping about it. The intensity was reflected in the unbelievably bizarre tactics of the contingency referred to locally as "the hospital side" and their willingness to risk the new hospital's success by using it as their primary staging area for their battle with Dr. Bailey.

Apparently in the '50s, like many small communities of that era, Waltham had one physician who rose to prominence and became a power to be reckoned with on the local political scene, particularly on the hospital board. That physician was the opinionated and outspoken, Dr. Bailey. Given his personality and natural tendency to lead (sometimes, dictate), it's not surprising that he created a small but dedicated number of enemies during the two decades at the top of his game.

After eight years in Waltham, it was evident to me that over the years the majority of thinkers and achievers spawned there, left the area for greener pastures, and the relative few who eventually did return after their educations, did not do so with altruistic intentions in mind. Educations completed, those in this last group returned to Waltham for the purpose of using their educational advantage for becoming big fish in a small pool. Their reunion with the small group of remaining local political big fish want-to-bees, was a match made in heaven. Unfortunately for Dr. Bailey, most of these want-to-bees had remained that because of him. In short, a new hospital was being built at a time when the county courthouse was bitterly anti-Dr. Bailey and they, along with their political big fish want-to-bees, intended a payback time for old Dr. Bailey's heavy handed rule of the old hospital's board. These circumstances translated for me and approximately twenty other physicians coming and going from the community over eight years time, into a nightmare of manipulation, reputation bashing, paranoia, and rancor in which it was impossible to practice medicine. It included near physical encounters during medical staff meetings, vulgar language, and for a short time, tape recorders attached to the phone lines of the hospital ER and ICU, all in an attempt by the hospital administration to come up with anything that might be used against Dr. Bailey or any of his clinic physicians. The backbone of the hospital side's game plan, it seemed to me, was to saturate the new hospital's medical staff with newly recruited physicians obligated to the hospital financially, so that in all professional committees within the medical staff they had the majority vote. Once they had the necessary numbers, they could then threaten Dr. Bailey's clinic members with accusations of professional wrongdoing and coerce them into resigning their hospital privileges. A testimony as to just how bitterly self serving was the hospital side's agenda, was the undeniable fact that on any given day about fifty percent of the hospital's patient census (and its revenue) were from Dr. Bailey patients!! This was due to the fact that Dr. Bailey had an extremely large practice and enjoyed a strong and widespread grass root support among average working families throughout the county. With no real population increase for a decade and no evidence that any of the other three long-time Waltham physicians planned on retiring, the message was clear for all new doctors being recruited for the purpose of this voting superiority by the two factions. We would be left to scrounge for the few undoctored patients in the county. And so, all the elements were already in place in little old Waltham in the early '80s for previewing the mechanisms that would contribute in the future, to the death of medicine *the profession*. Physician surplus, hospital management with non patient focused agendas, physicians hired by and dependent upon the hospital for their incomes, and extreme competition for a limited number of patients had prematurely predicted the circumstances to come with the health provider glut and big business management of our nation's hospitals in the future.

Dr. Phillips was the first of the new physicians to be introduced to me by the hospital administrator, probably since he represented the first successful recruit by the hospital team. He was to be hospital based since he was a board certified radiologist. He was described by the administrator as "a wonderful catch". After my first few minutes of conversation with Dr. Phillips, it became clear why. He was loud, gregarious, young, and more than well-indoctrinated with the hospital administration's attitude and prejudice toward Dr. Bailey and his clinic. He had long jet-black hair which he swept straight backward to hang over the collar of his shirt. With his black eyes and black goatee beard, he had the appearance of a person likely to be more at home in mid-town Manhattan than in Waltham, Kentucky, and when he spoke, his diction and accent supported this impression. I learned from him that his recruitment to Waltham had not been as difficult or the great triumph that Hiam would have liked everyone to believe. He had attended the same under graduate school as I, and that school was only 60 miles distant from Waltham. While in school there he met and married a lady from Waltham, and upon graduation he obtained his D.O. Degree and subsequently, finished a residency program in radiology. When his wife's family heard that the new hospital was trying to recruit a full time radiologist they quite naturally wanted their new son-in-law and daughter back in the community and had talked them into coming by and interviewing for the position. Typical of Hiam, the new physician he presented proudly to the public as one he personally had recruited was not a case of recruitment but a case of a new physician literally falling into his lap. Dr. Phillips' appearance and personality were completely out of sync with the country folks of Waltham. Finding him in Waltham was akin to encountering bagels and lochs on the local Dairy Queen menu. He informed me that since the Radiology Department was not yet up and running, he had been encouraged by the hospital to open a temporary office in town. He took their advice and as he put it, "was playing at being a GP" a few hours in the evening in order to have "a few bucks coming in". Once the radiology unit was up and running, he would be glad to turn over his "pretty decent cash-only, night-hours practice", as he called it, to me if I were interested.

As it turned out I did find myself in need of office space since the hospital preferred my waiting on the completion of its eight-physician office building adjacent to the hospital and currently only half way completed. This building was as ominous as anything regarding any hope of a decent practice in the future. Its overly ambitious square footage provided a constant physical reminder of the hospital administration's totally self-serving agenda. Only poor planning or the uncompromising pursuit of a goal unrelated to the community's medical needs could explain the rationale of recruiting doctors in numbers sufficient for filling a building of that size, in a town of only 4,000 people. Having prematurely committed myself by buying a home, hospital administration showed little interest in helping me find temporary offices, so once Dr. Phillips gave up his cash on the barrel head GP moonlighting, as he referred to it, we took over that office space. My office hours were the traditional ones, but unlike the opening of my first practice back in Illinois, the appointment book was barely even sprinkled with names on opening day.

First recruited, Dr. Phillips was also one of the first to leave. Fed up with the internal strife of the hospital and its medical staff, he left after only a couple of years. His radiological skills and services while he was there were satisfactory as far as I could determine. It was amusing and revealing to me to that during Dr Phillips's after-hours clinic experiment, he had been introduced by the hospital administration with much fanfare and the calculated description for the community of being the first "real specialist" we've had here in Waltham. Despite the fact that a more medically sophisticated population would have considered a radiological specialist as a definite negative for choice as a primary care physician, Hiam's advertising probably did garner Dr. Phillips a few early appointments in Waltham.

Waltham was isolated and medically unsophisticated, a fact which the devious hospital administration frequently took advantage of while pursuing their real agenda. While the demographics suited Hiam's needs well, for me they presented a problem. My practice style was based on current literature supported data, with patient participation in decision making and patient education. In my previous office in Levelton, Illinois this style had been a practice building winner. Patients there enjoyed participating in decision making when they were presented with current information and explanations. Additionally, they felt assured of my competence when my clinical decisions and explanations made sense to them from their own investigations and experiences. In Waltham, however, the patients' expectations were based entirely on what they had experienced from older local doctors over the years and remained unexpanded by any personal reading or other self explorations. Not only was my emphasis on teaching and discussion ill suited as a practice builder in Waltham, it often led to patients being dissatisfied.

A very simple example, but one of frequent occurrence, was the patient with a common cold or allergy demanding a "shot of penicillin". Typically, a mother would bring her child in at the height of the grass pollen season, and her presenting complaint would be,

"Well, John here always gets one real bad [summer cold] [sic], and it always just hangs on till I take him to the doctor and get him a big shot of penicillin to get rid of it." Totally unaffected by my pointing out the child's classic allergic shiners (darkness beneath the eyes of children with allergies) and allergic salute (the repeated swipes at his nose by an allergic child) or by my explanation of what antibiotics simply can and cannot do, mom generally would respond with,

"Well, all I know is that old Dr. Bob always said a big shot of penicillin was what he needed and he'd give him one and in a few days he'd always be better." I would then waste more effort in explaining what really happened was that by the time she decided the "cold" needed help, the pollen levels had diminished and consequently, so did the child's symptoms. The improvement, coincidentally coming about near the time of the "shot" being obtained, led her to mistakenly attribute the penicillin shot as the predominant factor in the patient's improvement. By this point in my educational attempt, the typical Waltham patient was either giving me one of those [I can't believe you're really a doctor smiles] or were already heading out the door to find a doctor that would give them a shot of penicillin. After several months of losing new patients in this manner I began to realize that if I were to remain competitive in Waltham, I was going to have to change my attitude about teaching and explaining and instead, provide the patients more of what they wanted. It was obvious that the injection was what these folks craved. After all, I had already sacrificed the quality of my practice to a degree, through compromising with a few of them by offering them oral penicillin. In these instances, rationalizing that even if I were giving them antibiotics when none was needed, at least with orally administered drugs it would be easier to discontinue the usage in the extremely unlikely event of an allergic reaction. This did not satisfy them either. It was the time honored, historically perpetuated "shot" they put their faith in, and nothing else would do. Putting personal and professional pride aside, I had to realistically face the economic facts here. A community with a fixed population, few of which were without personal physicians, suddenly finds itself inundated with new physicians to choose from, creates a buyer's market! I began considering other injectable medications that might be used for satisfying the shot mentality but with less potential for adverse effects than penicillin. The ideal substitute would be something that would at least help the symptoms, even if not affecting a cure. Injectable antihistamine/decongestants and corticosteroids fit the bill if used sparingly and judiciously. Using either of these two better assuaged my conscience that I was not giving them a placebo shot and charging for it. In the process of wrestling with these issues that had led to the lowering of my professional standards, a great revelation became evident which, projected into the future, could not bode well for the standards of medicine to come with its increased competitiveness and health provider glut. The sad fact of the matter was that the artificial doctor glut of Waltham enabled a group of medical consumers, ill prepared for recognizing quality medical practices, to dictate what they wanted from their physician and receive it. Simply put, supply and demand had enabled Waltham patients to force physicians into bad medical practices and lowered professional standards, just as it would nationally in the future. This lesson learned early in Waltham took on profound implications for predicting medical practice trends to be expected in the physician flush years of the '90s, and the picture wasn't positive. Medical practice in Waltham had begun to be driven by business principals rather than professional ones. Most of the country would not experience this trend for another fifteen years, but in the unique competitive environment there it arrived prematurely. Competition for patients in an over-doctored market opened the profession's doors to the-customer-is-always-right business philosophy, replacing the do-what-is-right-for-the-patient medical philosophy that had served an honorable profession and patients admirably for decades previously. And so nearly two decades before the rest of the nation, physicians' practices in Waltham became more askew from the principles taught in medical school. More prescriptions were written for sedatives, sleeping pills, diet pills, and pain pills. Useless antibiotic prescriptions became the standard for all *viral* illnesses. If a patient was intent on having a specific test done or referral made, it was always ordered, provided of course, there was money or insurance to pay for it. It mattered not that no medical reason supported the need for it. What mattered most was that the patient not become dissatisfied and leave the practice in search of a new physician who would provide him what he wanted.

Outside of Waltham in those days, competition had not yet compromised the profession to this degree. If a patient in most other areas of the country drove across town to a new physician for the purpose of obtaining a potentially unsafe diet pill which their own physician had refused to provide them, it was likely they would be met with another refusal and a little education as well.

"Well, Mr. Jones, there is a good reason why Dr. Sims didn't want to give you that diet pill. It isn't altogether safe for people with blood pressures in your range. Oh yes, probably you'd get by without a problem, but it's just not medically sound to take even a little risk just for taking off a few pounds for a while. I know Dr. Sims and he is a very good doctor. That's why he doesn't want to give it to you." In Waltham's contrived, over-doctored microcosm, however, practices became businesses prematurely, and survival replaced professional principals as the doctors' chief motivator. All of us surrendered to it in some manner, some much more than others. At the minimum, most gave in to the-customer-is-always-right philosophy, sacrificing their professional standards in the process. Maximally, others sacrificed their personal integrity by becoming yes-persons for the hospital administration, engaging in outright lying to their patients, and slandering of their colleagues. All this, due to physicians suddenly finding themselves with too few patients and a manipulative hospital administration constantly taking advantage of the anxiety and competition, while relentlessly pursuing their hidden agenda…the professional and financial ruin of old Dr. Bailey.

It was both extraordinary and revealing that a rural community of less than 5,000 persons in the early '80s could boast of a medical staff having four general surgeons, an oncology surgeon, a pediatrician, an orthopedist, a general internist, a board certified pulmonologist, three board certified family practitioners, three general practitioners, a urologist and a board certified neuroradiologist!! Doctor shortage, my aching stethoscope!

The neuroradiologist was definitely a diamond in the dime store. She was of excellent academic background and probably one of only a handful of physicians with such credentials in the entire country at that time. She was a highly professional, pleasant, and dedicated individual who would never have even slowed up in Waltham had it not been for the fact that her home town was only twenty miles from there, and, most significantly, that she was the wife of old Dr. Bailey's general practitioner son. She could have written her own ticket for any major urban hospital medical staff in the country, a fact she quickly proved after leaving Waltham. The departure of such a delightful, competent, and motivated individual at the hands of the hospital administration's continuous badgering and insults, speaks more clearly than anything to that group's true agenda and the ruthless effectiveness which they brought to it.

Such numbers and credentials of physicians for a rural hospital's medical staff in the early 1980's substantially corroborated what I had felt since serving as physician recruiter for the hospital board in Illinois. Despite the lay press's, the general-public's, and the medical schools' outcries, there was no dire doctor shortage in the early eighties. Waltham's unique physician supply was perpetuated by the hospital group's vendetta, and they were able to accomplish this by utilizing another premise learned in my brief tenure as physician recruiter on the Levelton hospital board. The Waltham hospital administration repeatedly took advantage of the mal-distribution of physicians and migratory patterns that I had first observed as physician recruiter in Levelton. That is to say, the most desirable living areas in urban environments were the most sought after by physicians for their living conditions. Periodically, when competition became tough, an overflow would occur, and the trickle would follow the path of least competition to the next most desirable living area. Despite media cries of a doctor shortage in the early 1980s, the Waltham hospital administration was able to utilize this physician migration pattern for keeping a constant replacement in the wings for any of their recruits who eventually figured out there was not enough work there for making it on their own or for those receiving their income directly from the hospital who decided to rock the hospital's boat by not supporting the hospital faction's Dr. Bailey mission. Unfortunately, this latter instance only counted for one or two over an entire eight year span. Once recruited by the hospital faction, these docs invariably ended up dependent for a significant portion of their incomes from the hospital in the form of guaranteed salaries, ER coverage pay, EKG reading pay, etc., etc. There was one physician, however, whose entire income was hospital dependent but who was an exception to the hospital's doctors usual profile. He was Dr. LaRue, the pathologist who had a contract with the new hospital since its doors first opened. He was a soft spoken, competent, and honorable individual whose periodic presence at medical staff committee meetings gave a temporary relief and normalcy to the otherwise contentious atmosphere usually present. Given that his integrity was obviously good, I wondered how he managed to hold onto that contract, since the clinical committees of the hospital were what Hiam and the hospital faction used for carrying out their bedevilment of Dr. Bailey and his clinic doctors. Initially an unrecognized commodity, I found myself appointed to some of these medical staff committees where cases were reviewed or quality issues were the focus, and it was always the same scenario. If the case had to do with Dr. Bailey or any of his clinic physicians, the hospital doctor faction would labor over every detail of the chart in the hopes of learning something with which they might build a case against competency or professionalism. In instances where nothing could be found (the majority of them), they would dig until they found some minor something, whether it be legibility of handwriting or unfinished paper work, which they would then place in the meeting minutes. The cases or issues concerning the hospital physician clique, however, were only superficially reviewed and assessed as satisfactory. My neutral colors quickly showed through on these committees, and I soon found myself persona non grata at committee meetings. I served long enough, however, for our pathologist to take note.

Thoroughly disgusted with the hospital administration and their obligated medical staff henchmen, I found myself seldom attending the regular medical staff meetings any longer. During dinner one evening I received an unexpected call from Dr. LaRue. He began by asking if I were planning on being at the medical staff meeting that night. Suspecting that he had been put up to the call by Hiam as a means of reminding me of my increasing absences at medical staff meetings, which the bylaws of the medical staff required me to attend, I was caught completely off guard when he took an entirely different tack.

"Well, I was hoping you would be there tonight," he said.

"I'm already wound down for the evening and have my jeans on," I replied, hoping he would take the hint. His voice now dropped an octave, and in an almost urgent tone he continued,

"Well, I really wish you would come on down. There are going to be some really important issues discussed that are very important to me, and I would appreciate your being there if there is any way for you to make it." Jokingly, I reminded him that my opinion didn't carry much weight with the hospital administration clique. "Well, it would be helpful to me, and I know you would understand what is coming down and would truthfully evaluate it," he said.

Five minutes later, still in jeans, I walked into the meeting room to surprised expressions of the hospital clique and to the smiles of Dr. Bailey's clinic docs. After some thirty minutes of the usual and boring statistical data, new business was reached on the agenda, and with that announcement came a pregnant pause which signaled I was about to learn why Dr. LaRue had wanted me there so badly. From the administration's physician ranks, Dr. Uhle, a pulmonologist recruited there by Hiam and lavished with every perk and advantage the hospital could provide for the several months he had been with them, spoke up. The gist of his new business was that he was not satisfied with our present pathologist (Dr. LaRue). His explanation for this was that the pathologist, to his way of thinking, was not experienced in reading bronchoscopy (procedure lung specialists use in which a scope is introduced into the bronchi of the lungs for obtaining biopsies or bacterial cultures) specimens. There followed a question or two from Dr. Bailey's clinic docs, asking for specifics. Then, a defensive statement or two from the pathologist at which point, unable any longer to complain in vague generalities, Dr. Uhle spat out,

"He never reads any positive findings."

"That's because there haven't been any yet," the pathologist calmly demurred. By meetings end I learned that in the nine months Dr. Uhle had been on staff, Dr. LaRue alleged Dr. Uhle had performed some 117 bronchoscopies without a single one turning up a positive pathologic finding. Old Dr. Bailey, in his inimitable manner, pointed out that in his 25 years of practicing medicine in that county he estimated the need for a patient to undergo bronchoscopy to have averaged less than two cases per year over that span of time. This was quickly and nastily countered by one of the administration docs with,

"Yes, that's because you probably missed diagnosing no telling how many lung cancers over your 25 years!" And so, the remainder of the meeting went like so many others I had attended in that hospital. We all had our say, and my own opinion was that the truth lay much closer to Dr. LaRue's statement than to Dr. Uhle's. The outcome, however, was determined before the meeting ever began, by the administration obligated physicians' preponderance on the committee. Dr. Bailey's statement, as thought provoking as it was, really only irritated them and fueled their determination. As usual, if he were against something, this group was automatically for it. Dr. LaRue's contract with the hospital was not renewed

Later that evening, in discussing the events with old Dr. Bailey, he expanded on his previous statement and recalled for me an earlier period of time when many of the nation's health policy gurus were touting the concept of hospital *regionalization* as the most economical and efficient use of the medical system's resources.

"As a practicing physician," Dr. Baily reminisced to me, "I didn't care much for someone limiting the services we could offer in our own local hospital back then. Now though, I am not so sure it wouldn't have been the right road to go down," he said.

In the regionalization concept, only certain hospitals would have been approved for say, OB services. So, if you lived in Waltham and were pregnant, there might be two hospitals within your geographic area that you could use for your labor and delivery. The obvious disadvantage was that both those hospitals might be located as far away as fifty miles. The advantages of such a system would be economically efficient allocation of technology dollars and a staff and a physical plant that were dedicated and designed specifically for delivering babies, a kind of specialty hospital, so to speak.

Dr. Bailey's comparison of the number of bronchoscopies done in our county prior to the arrival of the pulmonologist with the number done after his first nine months there, had profound implications for the patient/consumer when looked at from the hospital regionalization concept; again providing a preview of the potential problems that a subspecialties glut and business managed medicine of the future would hold in store for the consumer. Significant, but left unstated in Dr. Bailey's estimated average of two bronchoscopy referrals per year, was the fact that having had no pulmonologist in town in past times, he had referred those two cases to one of the state's two medical centers for the procedure. The only disadvantage in referring them was the inconvenience to the patient. The advantage to the patient, unknown to them but clearly understood within the profession, was that they were having the procedure performed by a physician accustomed to doing several such procedures a day. Additionally, the ancillary personnel assisting were also accustomed to caring for patients undergoing this particular procedure, and thus attuned to any and all problems or complications that might be associated with it. In the reality of medical practice, these two circumstances impact tremendously on the likelihood of a successful procedure, success being defined as obtaining the information or result desired from the procedure and with good patient outcome. This trend has held true in many studies and holds up for all areas of medicine. The likelihood of a good outcome is directly related to the number of times that a particular physician and hospital perform the procedure. This should not be surprising. If you made only one Italian cream cake per year, that cake likely would not be as easily accomplished or as tasty as if it were one of two Italian cream cakes you made every month. This is an extraordinarily important and valuable concept for today's consumers, given medical management corporations' business tactics of making small community hospitals compete for patients with larger, more medically sophisticated, and high-volume procedure hospitals in their locale.

Increasingly typical today, small hospital management will go after and recruit a surgical sub specialist such as a neurosurgeon, and from there, billboards and radio ads begin drumming it into the minds of the public that it is no longer necessary to suffer the inconvenience of a thirty minute drive to the tertiary care hospital in the city for their brain surgery! The unmentioned reality, however, is that medical specialties carrying out invasive procedures recommend a minimum-number of procedures per year that they feel a physician must be performing in order to maintain his or her skills. In the instance of a pulmonologist located in a small community such as ours, what did this mean? Clearly it strongly suggested that if he performed bronchoscopies only on patients of the community who had a bona fide medical need for the procedure being done, he would be doing very few procedures and with frequent, lengthy intervals between them. Just as clearly, his hospital support staff would not be specifically attuned to the procedure or complications associated with it, since a small community hospital's nursing staff wears many different hats and unlike the pulmonologist himself, did not come there with an extensive background and training for working with bronchoscopy patients. For Dr. Uhle this would mean that the longer he remained in Waltham, the more his skills would deteriorate from lack of use, were he to limit his bronchoscopies only to those patients actually in need of one. For the patient this would mean, that from an outcome standpoint, they would be better off travelling to a medical center for their bronchoscopy than having the convenience of a local pulmonologist, who was only infrequently performing the procedure, doing it in their local community hospital. Seldom getting to use his procedural skills, it was more than likely that Dr. Uhle or any other pulmonologist in such a situation would soon leave for greener pastures.

This is where Hiam and the hospital once more afforded me a preview of the medical system of the future, where business interests would call the shots economically instead of doctors calling them clinically. How often today do you pass through a small town within sixty miles of an urban center with a university hospital and see a bill board advertising that the small community's hospital now is doing coronary angioplasties (or some other subspecialty procedure) and urging you to take advantage of the "convenience" of having your procedure done without leaving your home town! Thank you, big business hospital management firms, for placing revenue interest above patient interest.

Hiam reassured Dr Uhle that, if he could hang in there long enough, he would become known to the physicians and hospitals of surrounding counties and eventually receive all their referrals for bronchoscopies. Never mind that the physicians of the surrounding counties were also aware that, statistically, their patients could expect better results by going to the centers rather than driving the same distance to our small hospital. Ignored also by Hiam was the fact that due to natural history and prevalence of diseases, for any pathological condition, there is a finite number of occurrences within any given stable population, and certainly neither Waltham nor surrounding towns were growing any. This meant that even if Dr. Uhle were to be referred every needed bronchoscopy from adjacent counties, he would still not be performing enough to maintain his skills or his livelihood. Determined to protect their investment, Hiam and the hospital began publicizing Waltham's new *lung specialist* in every way possible. This was easily done in Waltham due to the isolation of the community and its medically unsophisticated populous. The end result was that anyone, coughing from a viral upper respiratory infection for more than one day, sought out the "lung specialist". Hiam's other hospital faction doctors did their part as well, by asking for consults from Dr. Uhle for every routine pneumonia and moderate asthma case they admitted to the hospital. For patients this meant that they would receive standard treatment from Dr. Uhle for their routine illnesses at about three times the cost had they seen their personal physician for identical treatment. For the hospital the result was a win/win situation. They benefited from the publicity of having the only pulmonologist outside the major urban areas of the state, and they benefited financially from all the in-patient bronchoscopies he performed, the outcomes of the latter which had resulted in the called-meeting and its predetermined outcome that had sealed the fate of our honest pathologist.

Another of the doctors on the hospital's fast track was Dr. Ignacio. Dr. Ignacio had originally come to Waltham recruited by Dr. Bailey's clinic, but after several months there, he changed his allegiance over to the hospital forces. He likely was licensed via the FMG (foreign medical graduate) pathway and had grown accustomed to the urban environment of New York City while finishing his U.S. training there. By dress, mannerisms, speech, recreational interests, or any other parameter, there was no way that this individual chose Waltham as his first choice for locating his medical practice. Nearly all of his leisure time in the several years I knew him was spent in neighboring cities. He was an internist and endocrinologist, which is to say, that is how he was introduced to me by Hiam. Upon receiving him into the hospital's camp, Hiam began referring to him whenever talking with the town folks, as a "blood pressure and diabetes specialist". He also saw to it that Dr. Ignacio got to share rotation with the two internists from a neighboring community hospital and to receive payment, for reading our hospital's EKG's. This latter fact caused some contention between an ER nurse and myself when Dr. Ignacio read one of my four-month-old patient's EKG with a sinus tachycardia of 140 per minute as a "supra ventricular tachycardia"(in adults, an abnormal heart rhythm requiring urgent treatment). In very young children or infants, however, such a rate can be normal when crying or upset. The nurses in this new hospital were, for the most part, newly trained and inexperienced as well. This fact did not stop a particular nurse, however, from expressing her opinion on my lack of concern for the EKG diagnosis rendered by Dr. Ignacio. It did little to change her attitude when I pointed out for her that four-month-olds running temperatures of 102 degrees while screaming and upset at their surroundings, are not the same thing as a calm adult with a 140 heart rate. My explanation's lack of success may have been due to her minimal training and inexperience or to the fact that the nurses there, like the physicians, were usually either pro-hospital or pro-doctor Dr. Bailey's clinic, and this particular one was known for being in the hospital's column. I suspect it was due to a bit of both, as this was often the case and played right into Hiam's and the hospital faction's hands. The hospital faction capitalized on influencing an inexperienced nursing staff whenever possible for detracting from Dr. Bailey's clinic doctors or for making the hospital's doctors look better. In a different environment with experienced and objective nurses, Hiam and his cohorts would have had much less success in convincing them that black-was-white.

This EKG issue and Dr. Ignacio would later become even more enlightening with regard to the doctors mal-distribution and competition theory. Both being internists, the hospital faction thought it was quite natural for Dr. Ignacio and Dr. Uhle to go in together for the purpose of sharing call. Aside from the issue concerning the numbers of bronchoscopies without positive findings, Dr. Uhle seemed otherwise quite knowledgeable, not only in his sub-specialty but in general medicine as well. This fact was not lost on his new partner, Dr. Ignacio, or particularly appreciated by the latter either.

During this time frame I attended a medical staff meeting and learned that money had been approved for cardiac stress testing in our hospital. This was one of the few things that Hiam had actually shot straight with me on when I told him before committing to Waltham that I was currently doing stress testing in my own hospital in Illinois and would like to continue to do so in Waltham. He did indicate to me, at that time, that stress testing was not presently available in Waltham but was certainly in the planning for the very near future. The announcement that the equipment was now available was followed quickly by Hiam suggesting that Dr. Uhle and Dr. Ignacio set up some guidelines for its use. Dr. Ignacio then immediately made the motion that only the general internists on staff could conduct the actual test, and additionally, that an internist must be consulted about the advisability of ordering such tests. Not only was I not going to be able to conduct the tests for my own patients, they were trying to make consulting one of the two of them necessary for even ordering one. This met with an objection from me and from several other non-hospital allied physicians. Considerable discussion then ensued, eventually resulting in what the hospital faction described as "reasonable" qualifications for performing stress EKG testing. Numbers in their favor, the hospital enclave finally pushed it through that only a physician having completed an internal medicine residency and providing a letter from the cardiologist mentor of his training program of his/her competency in stress testing, would be permitted to perform the test. They were, however, unable to entirely monopolize the new testing equipment as they had originally intended, by making it mandatory for consulting Dr. Ignacio or Dr. Uhle before even ordering the test, much to the latter two's financial chagrin. Coming from a more medically sophisticated and larger community where I had hospital privileges for performing my own stress tests, this was certainly galling for me, particularly when remembering Dr. Ignacio's "supra ventricular tachycardia" diagnosis on the EKG of my four-month old pediatric patient. A few months later, passing a conference room door in the hospital, I noticed Hiam, Dr. Ignacio, and Dr. Uhle in a conspiratorial attitude at one end of a long conference table. I paused to glance at them and noticed all conversation cease, and a scowl appear on Hiam's face, a grin on Dr. Uhle's, and sheepishness on Dr. Ignacio's.

"What's going on?" I asked by way of a greeting.

"Oh, we are still working out the rules and regulations for stress EKG privileges." Hiam mumbled.

"Huh, I thought you worked that out quickly enough at last month's medical staff meeting," I quipped and continued on by the door.

"Wait a minute. You might be able to help us some here if you have a minute." I heard Dr. Uhle's soft-spoken voice. Backtracking to the door, but not entering, I said,

"Well, I doubt that. It sounded to me that you were pretty clear in your intentions at the meeting." Now Hiam's face was displaying one of his snake oil smiles he consistently mistook for charming.

"Now we didn't mean anything against you or want to upset you," he purred.

"That's right. We are just trying to prevent anyone conducting the procedure that doesn't know what they're doing," cooed Dr. Uhle.

"You know, like some of that bunch with Dr. Bailey's clinic," Hiam quickly added. Again from Dr. Uhle,

"I know you read EKG's well and I know you did them in the past and had privileges for doing stress testing at the hospital where you came from."

"Look, how would you propose rewriting the policy so that it would include you but eliminate the other bunch," Hiam bluntly interrupted.

As suspicious as surprised, I nevertheless, asked that they add to their present policy, "or any physician able to document a past history of privileging at another hospital."

"Great. Add it in," the first comment from Dr. Ignacio since I had entered the room.

"Thanks," I said and headed on down the hall wondering what was fishy here. Why would Hiam do anything for me? Rounds finished, I was in the parking lot when I ran into Dr. Uhle again.

"Come on. I'm not stupid you know. What was all that change of heart about in there?" He ducked his head a little and after a second raised it with one of his personable, impish grins and said,

"You know the way we had written it originally, that only someone who had gone through an internal medicine residency and could provide documentation from the cardiologist of their program that they were trained for reading and performing stress EKG's?" Before I could answer his impish smile flowed into a broad grin with his voice breaking with laughter, "Well, Dr. Ignacio's cardiology program director never replied to his request!"

Laughing with my office staff after sharing with them Dr. Ignacio's rejection by his mentor story, I considered his plight perfect justice. He deserved that embarrassment for his lack of integrity and inflated ego. Hiam and his administration deserved it too, given the fact they had sung his praises to the county folks as a "diabetes and blood pressure specialist." Although sad for the folks in the county, I particularly enjoyed the thought of Hiam out touting the virtues of a *specialist* whose own cardiology mentor chose not to respond to his request for an endorsement.

As much satisfaction as that information from Dr. Uhle provided me, quite honestly, the whole story regarding Dr. Ignacio's cardiology mentor's non response to his request for a letter of endorsement, was probably not on the table. In the same time frame that the stress EKG scenario was playing out, I had run across an article in one of our professional publications exploring hard feelings and tensions between cardiologists and general internists in the larger New York City hospitals. It caught my eye as possibly related to my theory on competition and physician migration patterns. The gist of the article was that New York City cardiologists were pushing for no physician to have admitting privileges to the coronary care units there, except cardiologists. This, of course, prompted the general internists to protest quite vigorously, as well it should have. For understanding the relevance of this article, it is necessary to be old enough for recalling that in the '50s and '60s, the general internal medicine specialist was still regarded in most medical communities as the authority on matters concerned with the cardiovascular system. In most of the country's CCUs and ICUs they were still at the professional apex of the medical staff utilizing these areas. This was, of course, before cardiologists existed in any real numbers and long before the advent of critical care specialists. The New York metropolitan area, along with Florida and California, were geographic areas considered desirable by physicians and, so, became the first having physicians of all specialties in sufficient numbers for competition to rear its ugly head. Once such competition became painful enough, it transformed its focus from recruiting patients to carving out hospital turf for more effectively competing. It had precious little to do with competence and everything to do with prestige and greed. Had the total number of general internists and their sub-specialist cardiologists remained such that all had plenty to do, it's likely the system would have functioned efficiently as in the past, with both factions getting along, and the internists consulting with or referring to cardiology colleagues those cases which were unusually difficult, or they were not making sufficient progress with. As medical schools continued to increase enrollments and more students chose specialties and sub-specialties, eventually cardiologists were graduating with plans of settling in the New York area, only to find that there were too many of them there already. Their solution was to bill themselves as the best trained for caring for cardiac patients, a statement not untrue on the surface, especially if the emphasis is on the word trained. It's what the statement leaves unsaid that leads it to project the unjust implication that general internists and family practice physicians are not qualified or competent for caring for many cardiac cases. It was then, in this environment and time frame, that Dr. Ignacio had the misfortune of requesting an endorsement from his New York City cardiologist mentor, only to never have him reply. I had little sympathy for him though, since like his cardiologist mentor in New York, when placed in competition for limited patients, he resorted to instituting his cardiologist mentor's same professionally unjust battle plan against the physicians of Waltham. Waltham once again was previewing for me the competitive future of medicine, and it wasn't a pleasant vision.

In spite of the bizarre hospital situation and cutthroat medical staff, our practice slowly grew into a sustainable although, by no means, lucrative one. Blessed with an excellent office staff, we had as good a time as possible given the frequent reports back to us of bad-mouthing from the hospital faction or from the hospital administration. We tried to lessen the sting on these occasions by considering the source and in knowing the real truth. However, it was never completely relaxed in the office, and most troublesome of all was how effective Hiam and his cronies were at convincing the most unlikely people that night was really day, people such as neighbors and fellow church members. The success he had was not so much a credit to his manipulative skills, as it was a discredit to the area's demographics. His techniques would have had minimal impact, if any, on my former Levelton practice or any practice I have since been in. It took a Waltham's rancorous history and a Waltham way of thinking for so many people to be deceived and so many wronged.

There were many good people there though, and we saw many of them as our patients. It did seem, however, that much more often than in my previous Illinois practice, I found it difficult to bring to our Waltham patients what was in their overall medical best interests. Sometimes this was due to misperceptions planted in them by Hiam and the hospital's propaganda machine, but just as often, it was due to naivete and lack of understanding of how the rest of the country utilized the medical system. One such patient was Mrs. Sarah. Mrs. Sarah was a widow in her late seventies who had little in the way of major health problems given her age. She would come in about every other month for having her blood pressure checked or whenever the occasional virus impacted her life. Although in good health for her years, this fact did little for her physical appearance as she was quite obese, and her facial features were several increments below the plain Jane category. One unfortunate day for both of us, Sarah came in to discuss having what she referred to as "this skin cancer" removed from adjacent to the bridge of her nose.

"Several years ago when I lived in Detroit I had one just like it removed by a dermatologist, so I know what it is, and I want you to refer me to someone to remove it," she informed me. Looking at this small actinic keratosis (a pre-malignant skin condition easily removed by a number of methods), I remember thinking, now here is an easy lesion for removal by anyone having available either cryotherapy or electrodesiccation. It was small and, given the advanced age of the patient, would never have time for developing into anything worse. However, not only was my office currently without cryotherapy or electrodesiccation, I had definitely not missed Sarah's presenting complaint, especially the part about *referring* her to someone to remove it. Obviously she had already decided that the task was beyond my abilities and training. I briefly considered offering her self-treatment instructions and a prescription for topical 5-FU (a creme form of a cell killing drug for external application) for topical self treatment of the lesion but quickly decided that, with my luck, she would either get some into her eye or panic at the initial inflammatory reaction and run off to some other local physician, in a state of hysteria. This latter possibility would have been no problem in my former Illinois practice. In fact, it would have been a practice builder, since whomever she chose to run to would reassure her that such treatment was perfectly acceptable, cost efficient, convenient, and that the initial local reaction was to be expected. She would then probably return to me for follow up and completion of the treatment, her confidence in me restored. Of course, in Illinois the doctors were not scrounging for new patients as they were in the artificial doctor glut created by Hiam and the hospital in Waltham. Secure in their practices and futures, the Illinois doctors would have been free to say to the patient exactly what should have been said in the patient's best interest. Today, it's the competition for the insured or financially endowed patient, that prevents physicians from always saying that which is in the patient's best interest or at least, being entirely truthful with them. The chances of Sarah being reassured by any other doctor she might run to in Waltham were slim to none, and I knew it. Given the dearth of undoctored patients in Waltham's doctor-eat-doctor environment, Sarah in such a scenario would simply represent too tempting a morsel for all the hungry practices there. Similar to undeveloped countries and their starving masses where it is hard to be concerned with integrity and philosophies when you're hungry, in over-doctored Waltham it was difficult for physicians to be concerned with character and professional integrity when their practices were hungry for patients. No, the thing for me to do was simply refer Sarah to one of two general surgeons whom I knew to be good, honest, and in no need of the business since their primary practices were in a slightly larger neighboring community. They were on staff at our hospital as a vestige of past times when Waltham had needed surgeons, and they had remained on staff there primarily as a service to their past Waltham patients from those times. Like our former pathologist, Dr. LaRue, these two docs' presence at our hospital medical staff meetings was like a breath of fresh air, bringing a semblance of a normal hospital's atmosphere for the short time they were there.

"Sarah, I'd like to send you to either Dr. Rohl or Dr. Patterson the next time they are in town, either of them will do a good job for you, and they are good and conscientious surgeons," I said.

"Oh, no, that won't work!" she quickly shot back.

"Why, have you had some bad experience or misunderstanding with either of them?"

"No, I don't know either of them," she said. "I'm sure they are both very good surgeons if you recommend them, but that's not what I need." Surprised and curious I could only ask,

"Well, what do you need?"

"Years ago when I had the first one of these removed in Detroit, it was a dermatologist who did it for me, and he told me then, if I ever got anymore of them, to be absolutely sure I went to a dermatologist for having them removed."

"But Sarah, I would have to refer you clear to Lexington or Louisville for a dermatologist, and you don't even drive."

"That's okay, I can get someone to drive me. I don't mind doing that, but they told me before to be sure and have a dermatologist do it." I tried explaining to her that way back then was before the advent of cryotherapy or electrodesiccation techniques that now made many physicians competent and capable of removing these lesions. Correction, I explained to the exam room walls, not to Sarah! She stubbornly clung to the anatomical-specialty prejudiced seed planted by that Detroit dermatologist so long ago. I reiterated the credentials and excellent records of the two general surgeons I had in mind, and for my added efforts got,

"No, it can't be a plain surgeon. If a surgeon does it, has to be a plastic surgeon!" I simply could not believe the bizarreness of the situation. Here we were with a total of five general surgeons on staff at our small community hospital, and this little lady wanted to drive eighty miles to have a procedure done that would have been unchallenging for me, were my office properly equipped for it. I was coming dangerously close to reminding her that even in her youth, bless her heart, a small scar should have been the least of her cosmetic concerns. Instead, I did the next worse thing. I lost my temper, and in a voice several octaves higher than usual, I heard myself suggest to her that she find some other doctor for misusing the medical system's resources with. I never saw Sarah as a patient again. My experience with her, however, has been reflected upon many times over the years as proof and example of another great contributor to the inflated cost of medical care, the uninformed or misinformed consumer self-referral. Left to their own devices, these individuals invariably ascribe to what I refer to as the *anatomical specialty* philosophy. If you have a skin rash you seek out a dermatologist, unless of course the skin having the rash happens to be covering your genitalia, in which case, a urologist or gynecologist might get consulted. At very best outcome, patients using this method of self-referral receive adequate care but at much higher cost and at much greater inconvenience to themselves.

In the past, when physicians of all types were scarce, those in specialties would not even see patients except by referral from another physician. This was for preventing their being bored by the more mundane of maladies and for making the most of their limited numbers by focusing their expertise where it was needed most. Should a routine medical case have been referred to the specialist of the past, he/she was not only bored but also insulted at the waste of their expertise. As time went on and physicians, including specialists, became more plentiful, the by physician-referral-only method was abandoned, and a specialist seeing an occasional routine illness simply put up with being bored. By the physician saturated '90s, encouraged by managed care's penchant for collecting specialty rates over primary care rates, specialists were still bored but never insulted, and glad to have the business provided by routine medical cases falling within their fields. Unfortunately though, unless U.S. medical schools revisit the prevalence data for disease entities truly requiring specialty expertise for successful outcomes and limit their specialty residency programs accordingly, the very near future could find specialists spending their days slogging through one routine matter after another, only to be shocked when the occasional bona fide case genuinely in need of their specialty expertise, demonstrates for them that their specialty knowledge base and specialty skills have deteriorated from lack of proper challenges.

Once in a discussion with a patient regarding efficient use of the medical specialties, the question was posed to me, "So, what if I'm a private pay and money is no object to me? What's the harm if I make myself an appointment with a cardiologist?"

"None at all, if your self-diagnosed problem actually turns out to be with your heart and if the cardiologist you select is a good cardiologist", I answered but reminded him, that even if a patient is savvy enough for understanding the intricacies of medical credentials and ends up making an appointment with a well trained, well credentialed cardiologist; good credentials are not always synonymous with good physicians. In addition to good credentials, the formula for a good doctor must also include judgment, character, communication skills, integrity, and focus. A wonderfully credentialed cardiologist who has a substance abuse problem is not the doctor you want calling in orders for you to a coronary care unit in the middle of the night. Likewise, a wonderfully credentialed cardiologist whose medical practice has been displaced by daytime stock trading as his passion in life, is not the one for formulating and communicating to you an intensely thought out management plan for your unstable angina. Nor should you want as your cardiologist, the eager, newly trained cardiologist just recruited by corporative management of the small community hospital for the purpose of propelling the community hospital into the coronary angiography business, especially if he lacks character and integrity. A cardiologist in the latter circumstances would be putty in the hands of individuals like Hiam and hospitals like Waltham's, with their special agendas. The patient might be inundated with advertising and public relation campaigns touting the presence and convenience of the hospital's new coronary diagnostic service, but that patient likely would never hear from a cardiologist of no integrity that a cardiologist and new equipment do not a quality coronary diagnostic program make. The point is that a patient would likely never voluntarily be told by a cardiologist of no integrity, intent on launching a maiden interventional cardiology service for a rural community hospital, that all newly initiated invasive programs generally have significantly higher rates of morbidity and mortality than their well-established larger and higher volume counter parts. Only a well-trained physician of integrity looking out for the patient as a whole would be likely to provide the patient with this priceless service and information.

In fact, primary care physicians are increasingly being placed in the hot seat over expansions of their smaller community hospitals' services, as small hospital management attempts to compete with larger centers for patients by offering similar services. It is particularly difficult for the primary docs whose professional integrity remains intact. These brave souls, fewer in number with each passing year, still assume a responsibility for looking out for their patients' overall best medical interests. The litmus test for this shrinking group is refusing to refer their patients to a facility or physician that they themselves would not send their own families to. If the numbers of physicians of this type today are small, their opportunities for doing an excellent service for their patients are even smaller, given the HMOs and other managed care organizations' increasing success at cornering the markets. Like Hiam and the administration at Waltham's hospital, the agendas of health management organizations and hospital management corporations today are increasingly not concerned with what is best for the patient but with what's best for business.

Ironically, the one area in which managed care could have impacted patient care positively by saving the patients from themselves, is the one area that they have been forced to back away from. Continued utilization of the gatekeeper concept, if administered by qualified primary care physicians, free of pressures from business forces and free from the-customer-is-always-right demands of the medically unsophisticated consumer, would have resulted in a vastly more efficient use of medical resources, while economically and medically saving the self-referring patient from the gaps in his medical savvy. Over time, I have become convinced that self-referred medically unsophisticated patients contribute far more significantly to the inflationary spiral of health care costs than generally is acknowledged. They could not have had this much impact, however, without another major contributor to the problem— the media.

Constant media fascination with newest technology in the medical sciences has led to medicine's becoming a mainstay in all major network news programs and frequently, the topic of their focused specials as well. The same course has been pursued by newspapers and magazines. After all, not everyone has a personal interest in world economics or the space program, but every single person has a vested interest in their own health. Unfortunately, media has always favored exotic and proceduralist specialties for their attention.

Waltham did have a large and physically appropriate skilled care nursing facility. Unfortunately, it too was impacted by the all-consuming battle between Hiam's hospital group of physicians and Dr. Bailey and his clinic. Its administrator was a nurse from one of the old guard Waltham families with business interests and political roots in the community. I never really knew where her true allegiance lay in regard to the battle, but the battle affected the function of the nursing facility anyway. With physicians from both the battling sides aware that hospital staff and the general public looked to an individual doctor's hospital census figures as a kind of score card for how the battle was going; this provided physicians a vested interest in admitting as many patients to the new hospital as they could. This pack-the-hospital mentality by the community's physicians impacted the skilled care nursing home's function by its staff having become programmed, after years of the battle, that when any of their patients developed an acute illness, they expected that patient's physician to transfer them immediately out to the acute care new hospital. This was despite the fact that being a skilled care facility, they were quite capable of administering IV antibiotics and most of the other procedures necessary for caring for the majority of nonsurgical and non cardiac illnesses. They were capable of handling pneumonias, urinary tract infections, and a host of other conditions just as effectively, more conveniently, and much more economically than the acute care hospital. Additionally, no consideration was given to the patient's overall best interests and needs. This led to total-nursing-care patients with multiple end-stage organ system disease status and no quality of life for years and no expectation of any kind of medical intervention changing that fact, being moved from their familiar beds and away from familiar faces to the new hospital for frequently what were their last days of life.

Once again, the artificial medical environment of Waltham provided a glimpse into the future, where the same misuse of skilled care nursing facilities and disregard of patients' overall best interests would be even more common, but today they are driven by entirely different mechanisms. Today, physician convenience, fear of misinterpretation of the paper trail by state regulatory agencies and Medicare, and fear of inappropriate litigation are at work perpetuating the continued markedly inefficient misuse of skilled care nursing facilities through out our nation, at tremendous financial drain to already dwindling Medicare coffers.

In Waltham, I had no hidden agenda for placing skilled care nursing home patients in the acute care new hospital for conditions which could be treated just as effectively where they were, and initially I tried avoiding this practice. Two patients stand out in memory as testimony to my dismal lack of success. The first of these was Mr. Shelby, a total nursing care patient, non-communicative and without any meaningful quality of life for several years. Strokes had long ago robbed him of his dignity and any meaningful existence. When first taking over his case, I asked the nurse if the patient had any family. I was told, none that she knew of. A couple years later while making routine rounds, I asked another nurse the same question. She had more information than the previous one.

"Yes, he has some family out in Arizona, a nephew I believe according to his chart." Quickly she then added, "But I've been here five years and I've never known him to have called or even to have sent a card to the poor man on Christmas or on his birthdays."

Within weeks of that conversation I was called at the office and told that Mr. Shelby had spiked a temperature and was breathing hard.

"It looks like pneumonia", I was informed by the nurse at the skilled facility. The call came in the middle of a very busy day at the office, and I gave the nurse orders to place the patient on oxygen, use Tylenol for his temperature, and began an appropriate IV antibiotic. Additionally, I asked her to inform the out of state nephew of his uncle's illness, what I was doing for him, and that I would be seeing the patient immediately after office hours that day. Ordinarily I would have made this call to the family myself, but this was one hectic afternoon at the office. Retrospectively, that was a big mistake to make in Waltham since I had no idea which side of the table Mr. Shelby's nursing home nurse sat regarding the hospital battle and therefore, what she thought of me. Giving her the benefit of the doubt, I assumed she would inform the nephew exactly what I had asked of her and, most importantly, in the proper manner and perspective. Within the hour my office nurse was telling me that I had a long distance call from the nephew of Mr. Shelby. Picking up the receiver I quickly identified myself and in response heard,

"This is Bill Evans. I'm Sam Shelby's nephew, the only family he's got. The nursing home told me he's got pneumonia, but he is not in the hospital, and I'm calling to see what's going on." Well, so much for the benefit of the doubt. I made that mistake a lot in the few years I was in Waltham, giving credit where none was due. You were never safe from the poisoned attitudes generated by the all-consuming medical battle and the personal bitter feelings which it generated. I quickly explained to Mr. Evans his uncle's condition, including his suffering and lack of any life quality for the past several years and finished up with my plan for treating the pneumonia vigorously and for keeping his uncle comfortable. I also brought the golden rule into the conversation along with my opinion that my approach would be appreciated by his uncle.

"Well, let me tell you something now," he blasted back. "If that old man has pneumonia, he belongs in the hospital. Not only that, I want him to have whatever specialists he needs, and if he has pneumonia, he ought to be in the intensive care unit, too." My luck was consistent. Not only had I asked the wrong nurse to make the family call, she was able to relate her slanted and untruthful view to a shallow-thinking and guilt ridden family member, intent on making up for years of lack of concern for his uncle in what he feared might be his last days of opportunity. The *guilty family* syndrome is a common and frequently encountered one for physicians caring for end-of-life stage patients, and it contributes unbelievably to unnecessary patient suffering and wasted cost within the health care system, likely multiple millions of dollars annually. The surest cure for it is meaningful immediate torte reform, medical end-of-life education for the general public, and effective monitoring and punishment for physicians and nursing homes who continue to engage in futile end-of-life interventions for convenience or financial reasons. All of us involved with geriatric care have encountered this type of guilt laden family member. Interestingly, but not surprisingly, the family who has kept in close contact over the years with a loved one undergoing the transition from meaningful life to physiological existence, seems to have no difficulty in appreciating, understanding, and applying the golden rule under such circumstances. In a very real sense, Mr. Shelby would have been better off had he entered his final days with no family at all, as given the numbers of disreputable attorneys and the capricious natures of juries; it was impossible to just hang up on the nephew as I should have. So, we went for the needless trauma and hospitalization route for this poor man's final days.

An even more memorable patient involving the skilled care nursing home of Waltham involved an octogenarian, a completely bedfast and total nursing care patient. This man had had no meaningful interactions or communication with other patients or nursing staff for several years. I remember making rounds in that skilled care facility on the first occasion after his care had become my responsibility, the nurse answering my query about this poor patient with,

"Oh, he's been pretty much this way for several years." On rounds a few years later, I observed this poor man's further decline, with increasing skin problems, problems maintaining weight, and generally, deeper into that declining phase of physiological life which our marvelous old, sagacious general practitioners of past days would have referred to as the "terminal dwindles". These same astute physician predecessors used to, before the antibiotic era, refer to pneumonia as "the old man's friend". It was in this "dwindling" period of this patient's years that I was notified he had quit urinating. Some simple lab tests and multiple unsuccessful attempts to get a catheter in place, suggested he had bladder outlet obstruction due to an enlarged prostate gland. While I was contemplating what was in this patient's best interest overall, the director of the nurses from the nursing home called to let me know that if anything were to be done to help this patient it would be better to do it in the local hospital. After futile attempts at convincing her otherwise, it became clear to me that the facility felt threatened by a death in their population, as this made their state licensure and Medicare surveys much more frightening for them. When I asked why this should be, I was informed that they were fearful that licensure or the Medicare surveyors might misinterpret something and think that this patient should have been cared for in an acute care hospital. For the life of me I could not believe I was having this conversation with someone who was supposed to be part of the medical profession. I wanted desperately to ask her how many physicians would be with the O.I.G. surveyors or the Medicare survey team. Why though, I knew the answer as did she, and it would only have led to my next logical question

, "Well, who do you think is better qualified for deciding what is medically, personally, and ethically in the best interest of this patient?" I continued trying by explaining to her that studies show that in the neighborhood of eighty percent of Medicare dollars are spent the last weeks of Medicare patients' lives, futilely attempting to bypass nature's inevitable plan. Nothing however, would change her mind and the pressure was on. Begrudgingly, I made arrangements for the patient to be transferred to the acute care hospital to spend the last days of his life.

All's well that ends well, however, although it would take another octogenarian for assuring that it did. The morning after the patient's first night in the hospital, I entered his room to find a meticulously and spiffily dressed, frail and elderly female seated in the visitor's chair with the morning *Courier Journal* folded neatly upon her lap and her wire-rimmed spectacles resting on the news paper. I knew in an instant that she had just finished reading that paper and now seated with her eyes fixed on the man in the bed, was remembering past times and circumstances. I introduced myself and was flabbergasted to learn that she was the patient's sister. She apologized for, what to me, seemed an immaculate appearance. She explained that she had flown in early that morning and had arranged ground transportation for the 70 mile trip from the Louisville airport to Waltham. I learned that she was from out west, and although she did not visit her brother over the years, she knew too much of his medical history to not have been checking in regularly over the years with someone in that nursing home. She knew exactly her brother's lack of any quality of life for the past decade and listened attentively, frequently affirmatively nodding her head as I explained to her his present problem. When I'd finished, I gently suggested to her that if she wanted my personal opinion on the care indicated in the best interest of her brother, I would be glad to give it to her, reminding her that it might be possible that another physician might not see it in the same manner. I continued by explaining to her that there were things we could do that might prolong his existence for weeks or even for another year. I explained to her the options of a suprapubic catheter or transurethral prostatectomy. Her eyes twinkled as they came directly glued to my own, and she asked point blank,

"Is that what you would want if that were you lying there?"

"Absolutely not," I easily answered her.

"I just don't want the old man to suffer. You can give him something for his pain and discomfort can't you?" she asked, with new moisture in her eyes. I assured her I could and would do my best to keep him comfortable, thinking to myself that it would not be difficult since uremia, the term for high levels of BUN in the blood stream in kidney failure, usually brings with it a gradual decrease in the senses, level of consciousness, and ultimately a coma with death following. This is generally speculated to be a rather peaceful and easy death. There are professional journal articles suggesting that the ease of death by this mechanism is why feeding tubes and IV fluids should not be routinely and indiscriminately utilized in the terminal days of life, with the mistaken idea that a painful death by starvation is being avoided. I had already satisfied the question of the golden rule in my own mind since in this case, the process of uremia was already well underway as part of the man's terminal illness.

Much more frequent examples of death by uremia occur in nursing homes, often by patient self-induced uremia i.e., the patients deliberately quit eating and drinking. For many with cognitive abilities still intact, it can be their way of saying, "Enough is enough and I'm tired." It should be remembered that in this group, nutrition is the one thing they still have control over in their daily lives. For another group, probably the majority, insufficient oral intake to maintain renal function comes at the end of one or more preceding medical conditions, such as stroke or dementia. Unfortunately for the patients in the group still with cognitive function and making a conscious decision to quit eating and drinking, I've witnessed many instances where such patients' wishes were completely ignored, and *enteral* feedings (feedings administered per tube inserted through nose into stomach or via a surgically placed tube thru the abdominal wall into the stomach) begun anyway, with the result being that the patient is stripped of his last act of independence and suffers additional days to weeks of needless prolonged suffering. Sometimes, however, the patient's family, even after receiving a clear explanation of the end-of-life circumstances of their loved one, will still insist on a feeding tube. This is understandable, but physicians should make every effort at making families in such circumstances realize that feeding tubes in the last days of life are a futile and selfish attempt to prevent the inevitable and a needless burden and source of additional suffering for the patient. There is never a right time for giving up loved ones, and in such emotional circumstances the golden rule is not as easily applied. My experiences have been, however, that where the family has been in frequent visitation with the patient, they know all too well the difference between physiological life and quality life, and usually elect to maintain their loved ones' dignity and comfort without the invasive tube feedings.

In both categories of end-of-life patients, the cognitive deliberately refusing adequate nutrition and the non-cognitive whose nutritional deficit results from the natural course of multi organ system end-stage-disease status, pressure from patients' families pales in comparison to the pressure the attending physician encounters from nurses and other care giving staff. Not only do nursing staff share in-common the well-intentioned but misguided reasoning of some patients families, this group also is dealing with the three fears: fear of adverse publicity for the facility, fear of L & R and other regulatory surveyors misunderstanding, and fear of legal repercussions from the ambulance chasing contingent of our legal profession. In fact, over the years I have found concerned families of patients who have maintained contact and followed their loved one's decline from meaningful life to physiological life, much more capable of arriving at a decision in the best overall interest of their loved one than caregivers as a whole. Care givers always maintain a degree of defensive paranoia that they will be blamed for something, leading them often to act in their self-interests, rather than the patient's best interest.

Little old Mrs. Jarvis at her brother's bedside, however, was as wise as she was brave and caring. Once the issue of her brother's physical comfort was settled, she began reminiscing about her brother's younger and happier years, painting me a picture of an independent, extremely physical, and vibrant individual. I couldn't help but notice and be amused at the frequency with which she referred to him as "the old man". After all, both their heads of hair were hoary, and both their skins were finely crinkled. Finally, she, paused briefly and looked up at me for my reaction.

"Mrs. Jarvis," I gently queried, "would you think me of terribly bad manners to ask your age?"

"No," she pertly replied, not at all, I'm 83!!" Her brother, "the old man", was 82! Simply and elegantly, Mrs. Jarvis had addressed, out of common sense and love, a geriatric concept being ignored through out our nation's hospitals and nursing homes: meaningful, quality life vs. physiological life. Due to not understanding the difference between physiological life and meaningful life or realizing the premium which the majority of elderly patients place on the latter and the fear they have of the former, nursing home patients in their last weeks to months of life are being put through additional days to weeks of needless and unnecessary suffering via futile medical interventions, and Medicare is being charged multimillions of dollars in a vain attempt at preventing the inevitable.

Of all my experiences in the surreal medical atmosphere of little old Waltham, none bode more ill or had more implication for the sad future awaiting medicine, than the continuous demonstration by Hiam and his supporters that an abundance of individuals licensed as doctors existed and the reactions of those physicians when forced to deal with competition.

Hiam and the hospital had a tried-and-true formula for maintaining high profile publicity and for deceiving the county's population into thinking they were on the verge of becoming a regional medical center in the state. Their formula relied on an inexhaustible supply of physicians being available for recruitment, a gullible populous, and the rally cry down at the courthouse of, "Let me tell you what Dr. Bailey and his clinic did today." Hiam always resorted to this latter technique on the rare occasion that a hospital board member or politically wired courthouse crone bothered to question any of his decisions or methods. This diversionary tactic was extremely effective for him. The animosity of his cronies and board was so great toward Dr. Bailey, that after they had cussed and fussed about Dr. Bailey and his clinic for an hour or so, they never got back around to asking Hiam the original question. No consideration was given to the effect on patients utilizing the hospital or recruited doctors' families and careers. They were considered entirely expendable, again providing a glimpse into the future of physicians and patients under a business driven health care system.

I remember attending one hospital meeting being held in regard to future physician recruitment goals, being motivated to do so by the local newspaper articles suggesting that Hiam and the hospital were considering another surgical specialist and another family practice physician for their next efforts. It turned out that their sights were set on a urologist as the specialist. This being the case, I had once again to appear as an adversary of the hospital faction. Regarding the family practice physician, I had no guilt whatsoever in explaining to them that two new primary care physicians had been in Waltham for over a year now, neither of which was as busy as desired and to date, none of the original primary care physicians in town had given any inkling of considering retirement. I offered all present the option of dropping by my office for a look at my accounts receivable and my appointment book for the last year. It was an easy sale for Hiam, however, given the fact that he had already managed to convince his hospital faction that I was pro Dr. Bailey clinic, rather than uncommitted and reflectively realistic. Making it even easier for him was the fact that the family practitioner they had in mind was a local boy from a very old and anti Dr. Bailey Waltham family.

Next, I turned my attention to the urologist, pointing out all the factors indicating this to be a poor choice for the latest specialist recruited for the purpose of marching for show before the public and media for the next year, or for however long the unsuspecting physician managed to remain in town. I explained quite earnestly and clearly why there would be insufficient urological cases for supporting a full-time urologist in Waltham. I went over the population numbers necessary for generating sufficient volumes for sustaining a urologist, pointing out that he or she would need consistently to draw from a minimum of four surrounding counties. I explained to them why this would not be possible since Shelbyburg, a neighboring town located only eighteen miles distance from our hospital, already had an established urologist of excellent reputation among patients and referring doctors alike. I also tried making them understand that the Shelbyburg urologist was only able to be there by being able to draw referrals from the surrounding four-county catchment's area. I implored them to consider what quality of urologist they were likely getting, if knowing these circumstances, he chose to come anyway. I asked them to remember their responsibilities, as board members and hospital administrator, for providing the community only the best quality doctor to be found, pointing out that once recruited and endorsed by them, some of their friends and neighbors were going to turn their health and well being over to this individual. I also asked them to consider the urologist's professional needs and the effect that an unsuccessful year or so in a strange community might have on his family. Not one of them took me up on my offer for reviewing my office ledger, and they all voted to recruit the urologist!

Once there and practicing I would occasionally run into the new urologist in the hospital. On several occasions I stood waiting for him to finish the charting he was working at in order to introduce myself. Each time, when finished, he would reel suddenly and walk away without ever looking up, so that our eyes didn't meet and he should not have to talk to me. It was clear to me that he looked at me as his detractor and just as clear, who was responsible for his indoctrination. The urologist left after only a year or so there, sickened by Hiam's nonsense and having an insufficient practice. Discussing that recruitment meeting with my office staff, all who had years of medical experience, they too were incredulous at the board's lack of insight.

"How long could they expect him to stay, since even if he performed a radical prostatectomy per week he'd run out of males in this county within three years," one of the nurses stated. A little over exaggerated, but it was an excellent point. I reminded her how the game had been played in the case of the pulmonologist. The hospital administrator and his hospital faction would provide an advertising blitz tailored specifically for the gullible, in which the urologist would be billed as a "specialist" of the urinary tract and encouraging anyone with a routine bladder infection to go and see him. They were very talented at this type of campaign. Before long I predicted, anyone with a bladder infection would be lined up in the new urinary tract specialist's waiting room to receive an antibiotic for their infection and paying three to four times the fee that their personal primary care physician would have charged for the same service. It was unfortunate, but while the rest of the nation was beginning to realize the advantages and value of having a primary care physician for overseeing their health care, the isolation of Waltham had left them stalled at the anatomical specialty mentality. After years of having the same five general practitioners adequately meet most of their health care needs and being referred to the larger centers for the remainder, these folks were enamoured with "specialists", and they mistakenly associated the term with up-to-date quality medical care. They had absolutely no concept as to training requirements, board certification, or, for that matter, even when specialty care was necessary.

Whenever attempting to educate patients on this matter, I would spin a hypothetical story for their consideration. I would ask them to pretend that a cardiologist shows up in Waltham tomorrow. He places an ad in the paper that his office is open for the practice of cardiology. Without any idea whatsoever of his qualifications or abilities, people in Waltham would run to his office with every chest discomfort they had. Some of the other doctors in Waltham might even have access to his credentials and know that although he completed his required years in a cardiology residency, he never was able to pass his boards or maybe, that he was considered less-than-average by his fellow cardiology residents, and he chose to not even take his cardiology boards, since he feared failing them. In both cases, however, he still legally could hang a shingle saying he was a board eligible cardiologist. At this point in the hypothetical teaching story I would invariably be interrupted by my patient with the protest,

"Yeah, but he still probably would know more about heart problems than a regular doctor, wouldn't he?"

"Maybe or maybe not," I would reply, explaining that individual doctors vary in their talents, motivation, and integrity. For the sake of learning, I would invite them to assume that the cardiologist's training and credentials were impeccable and, this being the case, consider if anything bad might result from the Waltham approach of anatomical specialist self-referral to such a cardiologist. Then I would describe the following hypothetical clinical scenario for their consideration.

A 70-year old male patient, with no history of cardiac problems, is struck by sudden sharp left-sided chest pain while struggling on a hot summer's day to pull up a deeply rooted weed. He comes into the house complaining of the pain and holding his left chest in a splinted manner for avoiding pain. His family is scared. They've never seen him in pain. He's 70-years old and male. They fear the worst.

"Come on, Dad. I saw in the paper that we have a new heart specialist in town. Let's be sure you're not having a heart attack." At the cardiologist's office, after listening briefly to the story, the cardiologist asks the patient to lean forward and twist his trunk to the left while stretching out his left arm.

"Ouch, I can't do that!" the patient says midway through the maneuver. Smiling, the cardiologist knows already that the patient is not having chest pain due to a heart attack. He reassures the patient and the family and settles back for taking a more detailed past history on this new patient. He learns that the patient has been extraordinarily free of any major medical problems. He's told that the patient has two brothers, one who is 58 and has problems with blood pressure and cholesterol and an older brother who is 78, who did suffer and survive a heart attack at age 73. Additionally, the patient admits that he was told by a doctor a few years back that his own cholesterol was too high and that he was given a diet and instructed to follow up with another test after three months, but he never did. He learns as well, that like most Waltham males, the patient has smoked two packs of cigarettes per day since age 15. Now the cardiologist questions the patient closely for any symptoms or signs suggesting heart pain in the past, particularly while working hard, walking up hills, or when angry or upset. There are none which he is suspicious of, and the cardiologist is now ninety-five percent certain, within his own little heart, that this man has not been experiencing any symptoms or signs of clinical heart disease to this point in his life. He is considering reassuring him and giving him samples of some aspirin-like non-steroidal anti-inflammatory medication for his intercostal muscle strain, caused by tugging at the root, and sending him on home. Suddenly, he has an afterthought that slows him up. This man is not having clinical symptoms or signs of heart disease, but he is, after all, 70-years old and statistically, he has to have some disease of the coronary artery system already present. It just isn't enough yet for him to have become symptomatic. He remembers as well the patient's brother who had a heart attack at age 73, just three years older than the patient himself. Now the cardiologist begins to enter the personal defensive mode. He considers what a disaster for a young cardiologist just starting out, if he were unlucky enough to send this patient home and within the next few days his asymptomatic but aged coronaries should become symptomatic, with the first symptom being a heart attack and death. Should this happen, his practice would be ruined before even getting started, and he might even find himself the defendant in a nonjustified lawsuit. He knows that this patient's family would never be convinced that this man's left sided chest pain of today's visit was musculoskeletal as he had just told them. They could never be convinced under such circumstances. Therefore, the cardiologist puts down the samples of anti-inflammatory medication and announces,

"I don't think the pain you are having today has anything at all to do with your heart, but every 70-year old man has some degree of coronary artery disease, and the only way we can be absolutely sure is with some tests."

At this juncture, the cardiologist is no longer treating the patient. He's now treating the patient's family and himself. Consequently, the patient winds up getting a resting EKG, a stress EKG, coronary arteriogram, or some combination thereof, depending on how much reassurance the cardiologist needs that he and his new practice are not going to fall victim to fickle fate with this type patient and his 70-year old coronary arteries. The result of such a hypothetical visit as this would most likely be that nothing would be found on testing that would significantly change the cardiologist's management of the patient. Of course, the costs of such testing could approach two or three thousand dollars, maybe more, and there is always the ever present chance of a medical complication occurring as a result of the coronary arteriography.

In Waltham, the patient and his family likely would leave the cardiologist's office pleased as punch and spreading the word throughout the county,

"When we thought dad was having a heart attack we took him to that new heart specialist, and boy that man is thorough! He did the latest tests just like they do in Louisville or Lexington, just to be sure it wasn't his heart. He isn't like those old doctors around here who probably would just have listened to him, poked on his chest, and sent him home with some aspirin!"

Just as the cardiologist knew within the first few seconds of history that this patient's chest pain was not related to his heart, a family physician could, with just a few more questions, treated the patient over the phone, saving the family the effort and cost associated with an office visit and with a statistically favorable likelihood of an excellent outcome. Like the cardiologist however, the age of the patient and the family's inability to understand the reasoning behind the diagnosis, would likely have made a family physician practice defensively just a little, by having them at least come in to the office for an exam. If after demonstrating to them the nature of the pain and its obvious association with the chest wall, the family doctor was convinced they understood the likelihood and reason this was not heart pain, the patient could then have been offered anti-inflammatory medication. Additionally, being attended to by his personal family physician, the patient has the advantage and safety factor of the family physician personally knowing his renal function status, allergy profile, and lack of ulcer history, all of which need to be known and considered in order to safely prescribe anti inflammatory medication. Having a personal long-term and mutually trusting relationship due to his role as the patient's family doctor, it would take far less testing for the family doctor to achieve his safety zone regarding fear of unjustified litigation, and, consequently, the proportion of the office visit fee charged by his family physician that would be attributable to *defensive medicine would* be minimal to none.

The public's lack of understanding concerning acceptable clinical practices and statistics has them seeking 100 percent accuracy in diagnosis and assurance of a good outcome in every instance, when in reality medical science is not always capable of this, nor is it even taught in medical schools that this is possible. The degree to which the patient is successful in projecting his/her unrealistic expectations to the doctor, and the willingness of the self-serving legal system to encourage the public's unrealistic expectations, combine to pressure physicians into an unreasonable and expensive quest for infallibility thru adding on of needless additional tests and studies. The degree to which the physician pursues this impossible goal depends on how much reassurance he needs for feeling safe. It's a personal thing and different for each doctor, depending on individual personality, circumstances, and depth of his/her training. As managed care continues to wipe out traditional long term trusting patient-doctor relationships, it's certain that most doctors will pursue their own safe-feeling levels by ordering up even more needless testing than ever before.

With medical management corporations' current fallacious thinking in today's hospitals that complex paper trails are litigation protective, the higher we can expect health care costs to spiral, particularly since the major input into the formulation of these paper trails are from bureaucratic agencies, administrative persons, attorneys, academic-world physicians, and non physician paramedical people. This leads to the policies themselves reflecting nothing of the concept of statistically acceptable clinical judgment as traditionally taught in medical schools. Instead, they are bloated with additional and frequently, unneeded tests and studies whose purpose is reproducibility within the paper trail, rather than any positive impact on patient care or case outcome. A typical example of this would be the pharmacist conducting Medicare required reviews of medications in the nursing home asking the attending physician to randomly and periodically check the hemoglobin and hematocrit on any patient receiving a nonsteroidal anti-inflammatory medication, since these drugs are known to have GI hemorrhage as side effects. Even if a hemoglobin and hematocrit on such patients were done monthly, patients still could have a life threatening acute bleed as the first and only bleeding event any day between the monthly tests, and the testing would have done nothing for preventing this. It does, however, cause needless discomfiture of a venapuncture for the patient and adds additional expense to the cost of care.

My years in Waltham found me continuously the victim of Hiam's and the hospital faction's well greased slander pipeline, all due to the fact that I chose to look at Dr. Bailey and his clinic with an open and unprejudiced mind. When the local newspaper ran an article on delinquent medical records and the dollars it was costing the hospital in collections, my name appeared prominently in the article. I came home that evening and, while working in my yard, was approached by my older next door neighbor, herself a long time resident of Waltham and a prominent business person, despite her naivete and simplistic way of thinking. She was a genuinely good person, but her simplistic nature made her rely on her old social acquaintances for her opinions and current event knowledge, even when they were plying her with gossip or deliberate misinformation. Smiling patronizingly and with the tone of voice of a mother letting her little boy know that she perfectly well knew the real truth, she cooed to me,

"Oh, Bob, I just wish so much that you would try and get along with Hiam and the hospital. You might as well since he is the administrator, and you could have an even bigger practice if you would just quit fighting with them. Like this not doing your records thing so that the hospital can't bill, it just makes you look bad when I know you really aren't."

The truth in regard to the medical records newspaper article was that I had two charts that were two weeks overdue and three that were approaching one month overdue. Dr. Ignacio, however, the hospital's biggest yes-man of all in matters of publicity and medical staff voting against Dr. Bailey and his clinic, had volumes of undone medical records, many over six months overdue and representing in the neighborhood of 185,000 unbillable dollars to the hospital. Worst of all and most telling of all, he continued being paid by the hospital for reading EKG's, and his remuneration remained uninterrupted despite his undone charts. Yet the newspaper account gave my name much more prominent attention in the article, even though my undone records were better than average for most physicians utilizing a hospital and of no financial consequence to the hospital whatsoever.

Isolated from any medical peers in the usual sense, due to the competitive circumstances resulting from over recruitment by the hospital faction, I had no source for the ego sustaining warm fuzzies usually provided by one's medical colleagues on deserving occasions and for which all real physicians used to strive. I was forced instead to take my compliments in any form and place I found them. Perhaps the strangest and most paradoxical place of all in which I found one, was in Hiam and his family. It's absolutely true! I regularly saw his preschooler son, and although he and his wife might occasionally see hospital faction physicians which they had recruited, if it involved anything serious, they frequently obtained a second opinion from me. Indeed, one such second opinion by me saved Hiam's wife a needless total hysterectomy at the hands of a general surgeon recruited by Hiam. As often was their practice, they had come to see me for a *second opinion* after a general surgeon recruited by Hiam, had recommended Hiam's wife having a total hysterectomy. His recommendation for a total hysterectomy in a lady of her age for the particular problem she was experiencing seemed questionable to me. I referred them to a Louisville gynecologist who concurred with me, and no hysterectomy was done. For all its unpleasantness, my Waltham experience afforded me a glimpse into the future, where business agendas would lead hospital managements to publicly espouse one philosophy or to publicly endorse certain physicians in the interest of their hospitals' bottom lines , while choosing different courses entirely for meeting their own family's health care needs.

When I finally left Waltham, about one and half years after Hiam himself had gone, it was a lot like beating your head against a wall. It felt so good to quit. At first I was bitter and worse yet, distrustful of people. After being away from Waltham a few years, I realized that Waltham during that period was a unique aberration and not representative of most of the country, not even most of Kentucky, thank goodness. Several years later, I heard the battle in Waltham had finally ended when old Dr. Bailey died of congestive heart failure in his eighth decade. The hospital was left a shadow of its former economic self and a hollow shell medically, given its potential in the beginning. The major local players in the hospital fiasco, along with Waltham's old-guard politicians and its younger *political-want-to-be* individuals, were forced to look elsewhere for a fighting arena in which to showcase their egos and for maintaining their positions in the local pecking order. They didn't need to look far. They simply picked up where they had left off when the new hospital had temporarily offered a new and more exciting battle venue. They returned from the battlefield of the hospital back to the school system, where they and their forefathers had honed their fighting skills and where their great grandchildren will likely still be fighting in the future.

CHAPTER 4

It is from these experiences, as a practicing physician during the two decades in which the medical profession became symptomatic and eventually succumbed to death at the hands of big business interests, that my opinion as to what was lost with that death has been developed. Clearly, economic forces have contributed the most to medicine's demise through more and more groups wanting a piece of the revenue action with each passing year. A study done in the mid-nineties at a major mid-western medical college concluded that the number of non MD clinicians, a group including: nurse practitioners, mid-wives, physician assistants, chiropractors, optometrists, and podiatrists, among others, had doubled in only five years' time. The study's authors predicted a further increase of about 20 percent by the year 2001 and, by 2005, that there would be more chiropractors than general internist MDs and more physician assistants than general-pediatrician MDs. At fees not dissimilar to those of MDs, what must the effect of this be on health care costs or on Medicare's and Medicaid's coffers?

If not thriving as a profession today, medicine certainly still thrives as a business, and the overhead of that business is anything but healing for the consumer. For understanding medicine's economics today, the consumer needs knowledge of its economics when it was still a profession along with an understanding of the factors impacting medicine as a business today. Particularly, the consumer needs to understand his own unwitting contribution to the problem.

The economics of medicine has always possessed inherent aberrations when compared to general-business economics, perhaps, because medicine was not a business but a profession, for all but its most recent history. The defining characteristic of that profession was the lengthy, expensive, and extreme educational requirements necessary for joining. Achieving the degree of MD requires eight grueling, post-high-school educational years and, depending upon the specialty pursued, another three to four years in specialty residency training. Most finish these training years emotionally spent and $50,000 to $200,000 in debt, to say nothing of lost earning power for their past ten years spent in school rather than in gainful employment. One article appearing in a professional medical journal in the nineties suggested that in the current business managed medical environment today, medical students, relying entirely on borrowed money for their schooling, from this point forward may find they will not be able to pay off their educational debts in their practices' lifetimes. For spending a dozen years of their lives without meaningful income, accumulating huge educational debts, and undergoing the harsh academic and emotional rigors of a traditional medical education, there is a definite minimal income necessary in most physicians' minds for justifying this arduous educational process. This is simply a common-sense fact and also the reason behind one of the most singular peculiarities of medical economics compared to economics in general. That is, in medicine, supply-and-demand does not necessarily apply. A small rural community having one physician typically might find that physician's office fee to be $50.00. A neighboring larger town with four physicians would support an average office fee of $60.00, and a nearby city with numerous physicians would likely have an average office call of $80.00. This is because all physicians want pay commensurate with their extreme educational requirements, but the overhead is much higher in the larger metropolitan areas. This deviation from the usual relationship of supply and demand to cost of the product or service is in stark contrast to business economics in general, which finds the consumer often driving to the city for a better automobile price due to higher volume and competition in the cities. Paradoxically, if this odd fact of medical economics should ever change, it will be a sad day for patients, since it could only come about via an extreme decline in the quality of the educational pathway of physicians. The reason being that there are no shortcuts to be taken en route to a quality medical education as traditionally taught in U.S. medical schools. Substitutes already are and undoubtedly will continue to be foisted on patients in the form of physician assistants, nurse practitioners, expanded paramedical provider privileges, and physicians imported from medical schools of questionable reputations. The oft quoted saw, *you get what you pay for*, unfortunately, has not held true in the cases of these physician-substitutes so prevalent on the health care scene today, as the fees of these individuals are now pushing at those of MDs, and this is bad news for the health insurance industry and for Medicaid and Medicare.

Another of medicine's economic oddities often debated within the business is the highly skewed reimbursement system in favor of surgery or procedures. Simply put, this means physicians performing surgery or a procedure of some type, regardless of complexity, are paid disproportionately higher fees than physicians not performing them. My best guess is that this inequitable system had its historic roots in the dark ages when MDs were scarce, and MDs with additional surgery training were as rare as hen's teeth. The great mystery today, with some sources reporting an oversupply of nearly every type of physician, is how this skewed reimbursement system has been perpetuated.

An example demonstrating how unfair this system is would be a case in which a primary care physician is managing the care of an elderly and critically ill patient with diabetic ketoacidosis in a community hospital ICU. That physician might well be up most of the night evaluating returning lab values and rewriting IV fluid orders as the patient's condition fluctuates. Likely as not, during the day that doctor would even have to cut back on the number of patients seen in his office in order to focus on this constantly changing complex case and for keeping in close communication with the ICU staff. In the end, he/she submits the bill for services rendered and receives from the insurance carrier a dismally inadequate payment— entirely disproportionate for his effort, time, and lost office revenue. That physician's partner, however, might perform two ingrown toenail excisions and two excisional skin biopsies, leave the office by early afternoon, and receive five-times the pay his partner received for managing a very complex and dynamic case over a seventy-two hour period of time. Certainly additional skills and training are required for performing minor surgeries and other procedures. Once learned, however, it takes no more thought or effort for carrying out some of these procedures than is required of a garage attendant for replacing a car's battery. A legitimate argument could be made that physicians doing surgery and procedures have much higher malpractice premiums to be recouped, but even adjusting for this there is probably still some fee-cutting room, particularly for the more routine surgeries and procedures. Today's technology has led to a myriad of procedures that are beneficiaries of this disproportionate reimbursement system, and there are probably very significant dollars to be saved for the consumer if the situation were to be addressed. Consumers today should not think this only of financial concern to physicians, as the number of technological procedures being performed and garnering higher fees has burgeoned over the last decade. Significantly higher reimbursements can be regularly claimed for performing a sigmoidoscopy, esphagogastroduodenoscopy, colonoscopy, bronchoscopy, nasolaryngoscopy, and colposcopy, to name just a few.

In the mid-nineties a professional medical journal featured an article exploring the profile of the typical medical student of the '90s. In the article students were being interviewed as to what specialty they intended pursuing upon graduation from medical school. One extremely candid individual essentially replied, unashamedly and matter-of-factly,

"I don't know just yet, but I want to be in one that has a procedure associated with it because that's where the real money is."

That young man's statement was appalling, alarming, and provides much insight as to what increasingly seems to motivate many of today's medical students in their decision to enter medical school in the first place and, clearly, points out the motivation behind their choices of specialties upon graduation. Recent press coverage has pointed out that the medical specialty residencies being pursued most vigorously by graduates of U.S. medical schools, exactly parallels the increasing average income of the specialties, with primary care receiving the fewest numbers of graduating students. Unlike past decades, those entering medical schools today are increasingly not doing so out of their passionate love of the basic medical sciences and their fascination with seeing these sciences applied in the diagnosis and treatment of illness. Just as clearly, most are not depending on honor and respect from their professional peers as a foundation for their self-worth and a measurement of the success of their practices. After all, the medical *profession no* longer exists except within the universities (I hope) and, possibly, still scattered in a few pockets throughout the nation as a whole, if you are lucky enough to find them. However, do not be surprised at some of the unexpected places you might still occasionally encounter a proud, effective, efficient, and caring medical profession reminiscent of the past or at the number of high-profile *big-name* venues where you will find its absence disturbingly conspicuous.

In the early '80s my office manager approached me and asked if I knew which patient's account she should apply a $150 check received from a health insurance carrier but having no specific patient account noted by them. When I did not, she contacted the insurance company's offices, and we were surprised to learn that the check issued by them was not for any particular patient account but for what they termed an "incentive bonus". They explained that it was for my contribution toward avoiding charges associated with unnecessary scheduling of inpatient surgeries. Since the only surgeries I did consisted of laceration suturing, excisional skin biopsies, and excision of ingrown toenails, we inquired further as to why I should be the beneficiary of such an incentive bonus. It was explained to us that the four general surgeons in our geographic area were also performing these same routine procedures, and their average physician charge was $150. Additionally, they almost always did their work in the local emergency rooms, the latter which, submitted further charges of $200 or more to the patient's insurance for usage and supplies. The sum and substance was that this made the average cost to the insurance company $350 for an ingrown toenail excision done by the general surgeons, while my charge for the same procedure was $75!! I took the office staff out to dinner with the incentive bonus, and, after thinking about it some, we all enjoyed a good laugh with the dinner. Typical of the physician ego, I had assumed we saw so many patients with ingrown toenails because word of mouth had spread on how effective and painless my work was. Yeah, right! It was word of mouth alright, word of mouth how cheap we were!! In truth, the reason I elected to do so many procedures in the office rather than the hospital ER was that the paper trail chase had already begun in the hospital by the early eighties, making hospital work inefficient and unpleasant. Today, I could no longer keep a $75 office charge for an ingrown toenail excision, since, thanks to continued bureaucratic meddling and continued growth of the paper trail demands, office overheads have skyrocketed. Even for participation in Medicare an office today has to have a computerized billing system and a computer savvy staff for its operation. This means higher overhead, and all shoppers know all too well who pays when overhead goes up!

The one legitimate factor in the equation for calculating the cost of healthcare today is the tremendously expensive technology advances of the last two decades. Their contribution to today's high-cost of medicine is straight forward to anyone who watches the evening news or reads the evening paper. Technology available today is astounding in nature, and its benefits to patient outcomes have bordered on miraculous— *when it is used appropriately, efficiently, and effectively.* Unheard of years ago CT scanning, magnetic resonance imaging, fiber optics, laser surgery, laparoscopic surgery, and thousands of new pharmaceuticals are readily available and accessible to physicians today. The bitter undeniable fact, however, is that such technology is incredibly expensive to research and develop and ultimately arrives on the market with a budget-busting price tag. The high price of today's technology is only part of the story, however. A surplus of healthcare providers for ordering this technology (very indiscriminate and inefficient ordering in the case of the non MDs now having permission for ordering them), the practice of defensive medicine, and an increasing irresponsibility among new-age physicians for ordering testing in an efficient and economical manner is ratcheting the cost of care skyward at a dizzying pace.

My first experience with high technology came in the mid-seventies when I ordered my first-ever CT scan. I remember well the patient who walked into my office in 1976 and told me he was there because his boss made him come in to find out why he slept so much at work. I asked him if he slept well at night and if he were tired when he first awoke in the mornings. He responded something to the effect that he just fell asleep easily during the day. I asked a number of other questions and examined him, focusing on the common causes for tiredness or fatigue such as anemia, hypothyroidism, chronic allergies, etc. Coming up with nothing, I told him we would do some blood work, and he should call back in a week.

"Okay," he said, reaching into his pocket for something. "Would you sign this for me so I can go back to work?" he asked. I took the paper which had the heading of a local trucking company and read the simple statement: I find no mental or physical illness that would prevent Mr. Jones from returning to his employment as a long distance truck driver for Ace Trucking, followed by a place for my signature.

"Would you mind my calling your boss?" I asked him. Five minutes conversation with the boss gave me an entirely different perspective on the nature of the young man's complaint. What the patient described as falling asleep easily and a lot, his boss described excitedly as,

"Doc, you would have to see it to believe it. We can be sitting talking, and the next second his eyes are closed, his head slumped to one side, and his breathing sounds deep like a snore." Returning to the exam room with his boss's description echoing in my ears, I suggested we get some further tests. Since CT scanning had just become available in the medical center nearest me, we scheduled him for a CT scan of the head. Late in the afternoon on that same day I received a phone call from a very excited neurosurgeon at the center, informing me that this unfortunate man had a large mass deep within the brain that appeared to be malignant.

"And he checked over so normal," I blurted into the phone.

"Yeah, he sure did. I happened to be present when he was signing in for the scan, and I thought to myself here goes another normal scan and some wasted time and money," my neurosurgeon colleague replied. This outwardly appearing youthful and happy man was dead within a year, despite everything the medical center attempted to do for him. Does that mean that every man presenting to a physician with a complaint of isolated fatigue or sleepiness should undergo a CT scan of the head? Of course not, it's in medical school and during their specialty residencies that MDs are taught the *appropriate and responsible* use of laboratory tests and imaging procedures, and none of our medical schools would condone such irresponsible use of such an expensive technology. All responsible physicians know this, or at least they did before defensive medicine entered the picture. Today though, it's more than likely that a physician having had such a rare experience as mine with the sleepy man and the brain tumor, would ever after order a CT scan of the head on every patient presenting to him with the complaint of "falling asleep all the time." The difference today is the extreme litigious atmosphere of the practice arena and the public's unrealistic expectations.

Although much more sophisticated and expensive, our modern day technology is merely an extension of our simple tests of the past. As such, its utilization should be governed by the same medical principle as any ancillary testing— for confirming or ruling out a suspected illness or condition *that the physician's history and physical have led him to strongly suspect.* This is not a new concept, but one that is traditionally taught in all medical schools throughout the U.S. It is a part of the reason that the many years of detailed scientific schooling are necessary for obtaining a license to practice medicine. Today, however, due to business pressures, cultural expectations, and defensive medicine physicians are abandoning their clinical thinking skills in favor of expensive high-tech imaging and testing for reassuring their patients and themselves. Additionally, under the influence of special interest groups, we are now also permitting non MDs to order sophisticated and expensive testing and imaging without any constraints or oversight, pushing health care costs ever higher.

Primary care nurse practitioners today are responsible for a steadily increasing percentage of the tests and procedures being ordered, and their effect on the cost of health care is much greater than the numbers of their ranks would predict. This is because there is an inverse relationship between the number of studies ordered for successfully managing a case and the amount and depth of education of the clinician ordering them.

For example, in evaluating headaches in the office setting, a seasoned neurologist might not order 8 CTs of the head per year. A very busy and rushed primary care physician would likely double the neurologist's number, and a slower-paced primary care physician, with more expertise and confidence in his/her neurological exam, would order more than the neurologist but many fewer than their neurologically weak primary care colleague. Left unsupervised, dollars spent for studies ordered in a suspected appendicitis case by a third year medical student will be 3 to 4 times the amount spent for an appendicitis work-up by that same student's third surgical residency year. The more medically learned the clinician, the fewer tests will be utilized. Besides adding additional hands into the already over-crowded Medicare cookie jar, the hands of most physician extenders are very economically inefficient ones as well.

MDs are taught in medical school that before any test, regardless of its price, is ordered, a doctor should have in his or her mind, as a result of their history and physical exam, a condition or disease that can be helped or at least which the patient has need to know of. Can you imagine going to your doctor's office because you haven't felt your usual self but unable to pinpoint any particular symptom for your doctor to work from, his telling you,

"Well, Mrs. Smith, in that case we'll order a complete blood profile, chest x-ray, EKG, stool studies for parasites and a CT scan of the abdomen and head!"

Increasingly today, because of doctors not following their medical training and another group of individuals being permitted to order testing without ever having even been to medical school, thousands of needless tests and procedures are being ordered daily. Tests should only be ordered for the purpose of proving or disproving a physician's diagnosis and nothing more. Adhering to this simple basic principle as taught in medical school would reduce the cost of health care in this nation by an unimaginable amount.

Efficiency and fiscal responsibility necessitate that a good physician always asks himself one additional question before ordering ancillary lab or x-ray studies. If the test ordered confirms the diagnosis, will it enable me to treat or make something better for the patient? In other words, why subject the patient to the expense and whatever risks associated with the test (and there nearly always are risks), if the results cannot be applied in the patient's best interest. This seems inherently simple, and yet, every day there are thousands of tests and procedures being done when they should not be, representing very poor quality medical practice indeed. Sometimes this is the fault of the physician, sometimes the fault of the patient, sometimes a combination of both, and very frequently, the media and defensive medicine play a role as well.

An illustration of this would be a middle-aged male with classic signs and symptoms of coronary artery disease and of perfectly sound mind, saying to his physician,

"You know doctor, I know you are right. I have angina pectoris just like my dad did. I remember watching him for years with it. I also remember him dying during his coronary bypass surgery. I'll take as much medicine as you like, but I've already told my family that nothing you or they can say will make me decide to have bypass surgery or coronary angioplasty."

In a situation such as this it would be a waste of money and time to suggest obtaining coronary arteriograms for this patient, to say nothing of the needless risk associated with putting a patient through a coronary arteriogram when nothing was going to be done for him on the basis of the findings. Today though, physicians push harder for tests of no benefit in management of cases due to defensive medicine, fear of the paper trail appearing too thin, and, all too often, just because the patient expects and wants them.

Many more typical examples of needless testing occur among the nation's nursing home population. Usually this involves an extremely elderly patient with no quality of life or expectation of medical intervention, regardless off type, altering that fact. This would be the little old man or woman who is a total-nursing-care patient with no cognitive function and having multiple organ system end-stage disease. The patient's heart, kidneys, musculoskeletal system, and brain are simply worn out and beyond any meaningful help from medical intervention. This type patient is not capable of communicating or demonstrating any cognitive function and is dependent entirely on nursing staff for meeting his/her daily living needs such as bathing, eating, toileting and clothing. A typical example of how inappropriate testing comes about under such circumstances is for nursing staff to report some "possible" blood noticed in the patient's stool. In patients not in their end stages of life, blood in the bowel movement often requires an extensive workup, usually including colonoscopy for being sure the blood is not coming from an undiagnosed tumor of the GI tract. The nurse reporting the blood is aware of this fact and notifies the oncoming shift to please make certain the doctor is aware of the blood noticed in the patient's stool. All of the nursing staff is now aware of the blood issue, and, if any communication with the patient's family or guardian has transpired, the staff will have already notified the family of their personal expectations of what will likely be done for the patient. Now both nursing staff and the patient's family are primed and expecting the usual colonoscopy routine to be carried out or at least for *something* to be done! For a patient in this particular situation, however, before anything at all is done, quality medical practice dictates that first, the patient's legal guardian or family should be made aware of the situation and given a clear description by the doctor as to what would be done in a younger, healthy patient having traces of blood in their stool. In short, see if the guardian would want aggressive diagnostic intervention for ruling out the existence of a cancer in the colon, when, in this particular case, medical intervention, regardless of type, will not improve this patient's quality of life. Jumping ahead is often helpful in putting it all into perspective for the patient, family, or guardian. This can be helped along by asking them to consider, for sake of discussion, that if a colon cancer were definitely there, would they believe, given the patient's end-of-life condition due to pre-existent, chronic medical issues, it in the best interest of this patient to undergo surgery for its removal. Having engaged in many such discussions with families over the years, I've found that ninety-five percent of the time they make the correct decision and the one in the patient's overall best interest. Clearly, once the decision is made and agreed on by all for no intervention if a tumor is found, then no further testing of any type is warranted in an attempt to prove or disprove a tumor's presence. As simple and common-sense as this seems, the principle is often violated due to physician or facility fear that doing "nothing" will make the paper trail appear incomplete, leading to unwanted confrontation with OIG or Medicare at survey time. Thus, most of the time, some type of additional follow-up study will be ordered even though, once the decision for no further medical intervention for that particular problem has been made and agreed on by all parties, further studies directed at that problem or not needed or indicated.

When medicine was still a profession, the accountability for end-of-life tests and procedures lay entirely with the attending physician, patient, and the patient's family. These were the only parties that needed to be satisfied as to what constituted the patient's best interest. In past times there were no bureaucratic paper trail requirements being generated by individuals exemplifying that a little knowledge is a dangerous thing and the paper trail then being over-interpreted by facility administrators made paranoid by regulatory agencies' increasing demands for documentation from multiple medical disciplines. This is the status today though, and it has led the administrations of hospitals and nursing facilities to become engaged in the practice of medicine by way of developing policies. These policies are neither physician nor patient friendly and frequently impact the patient's overall best interests and the health care economy adversely. They are designed expressly for making regulatory agencies such as the Joint Commission, the Centers for Medicare & Medicaid Services, and State OIGs jobs easier to perform and in the mistaken belief that such policies will lessen the facility's exposure to adverse publicity or loss of licensure. In reality, such attitudes are responsible for tremendous additional unnecessary dollars being added on to the cost of care through needless testing and consultations, to say nothing of the needless discomfort and risk to the patients undergoing them.

Often overlooked, but another major contributor to needless testing, is the news media. For the past two decades, with each passing year, the news media have become increasingly entranced with medical science technology. Indeed, most TV news networks today have their own health segment of each broadcast, complete with physician analysts. Such segments are certainly of some educational value, particularly to physicians. Unfortunately, physicians listening to such analyses of the latest technology are the *only* individuals in the TV audience with sufficient background for interpreting the relevant from the hype. Making matters worse is that the media generally focus on latest technology. Often times the things being reported on are still experimental, limited to university settings, and unavailable for use by most physicians for years yet. The end result for the general public listening to such broadcasts is that they become aware of the names of the latest drugs, procedures, and equipment. A little knowledge is a dangerous thing, and this leaves them with the misperception that the advancements that they hear of in the news are in widespread use, of documented safety, necessary in every situation, and without economic consequences.

The impact of major drug companies being permitted to advertise their latest and most expensive drugs directly to patients thru the media has increased the cost of health care stupendously when, in many instances, older and less-expensive drugs would do just as well. The private-pay patient or patient with insurance demands the antihypertensive drug he just heard acclaimed on last night's evening news and gets it! Competition for the moneyed patient, created by the influx of MDs from all over the world flocking to the USA as the last hope for lucrative medical practices, have made the customer, if well heeled, always right. Today, the patients with the ability to pay usually get what they request, whether good or bad for them. When medicine was a profession and accountability rested totally with the physician, misinformed patients could simply be told the truth by their personal doctors. If they chose not to believe their own physician, a second opinion from another doctor would likely result in the same truthful answer. However, in today's environment, where medicine is a business and patients are regularly competed for, patients are able to exert much more pressure on their physicians, and this results in doctors being coerced into prescribing the non indicated tests or specific medications being requested by the patient, in an attempt to keep the patient from becoming unhappy and seeking out another managed care organization or another physician. In short, when medicine was still a profession physicians were only required to act in their patients' best overall health interests. In the business of medicine today, the-customer-is-always-right philosophy dominates.

Another example of how a little knowledge provided to the potential consumer by the media leads to needless testing and increased health care costs, is exemplified by the prostatic specific antigen (PSA), a blood test frequently getting TV and other media space these days. This is a blood test for the presence of a serum biomarker that can suggest the presence of cancer of the prostate gland under certain circumstances. The test is difficult to interpret and is of best use to physicians for following post treatment results of prostate cancer. It is fraught with potential for over-interpretation at lower levels and in the very elderly population. Most urologists and all good physicians of other specialties do not routinely order this test as a screen on extremely elderly men, especially elderly men with other significant illnesses likely to take their life soon. This is because prostatic cancer, in many instances, is so slow-growing as to make the very elderly male much more likely to die of another illness before the prostate cancer would become symptomatic. In the time frame immediately after the media's first blitz regarding the PSA test, I cannot tell you how many patients I had inquire about running the test on "granddad" who, totally incapacitated by multiple other illnesses, was already in the last months of his life. Nurses in skilled care nursing homes and within hospitals have pressured me to order this particular test on patients who would receive no benefit from the diagnosis of prostate cancer being made. It takes a much more in-depth medical background than most nurses possess, let alone the general public, for understanding the proper usage and limitations of this test. It's simply another instance demonstrating that, as training and education increase, the use of ancillary testing decreases. This is an extremely important concept, given that today, special interests and business trends are pushing for increased usage of so-called physician extenders, a varied group that, regardless how much they profess to the contrary, do not possess a minute-fraction of the in-depth medical education possessed by a properly trained MD from a traditional medical school.

Also, seldom being addressed by the media and universally ignored by the public on the rare occasions they are addressed, are availability and cost. These twin emotional nightmares have eluded satisfactory resolution by the best of our politicians, ethicists, and economists. Is it any wonder that a lady without health insurance and pregnant with a child that she has just been told is hydrocephalic by ultrasound, should not believe that intrauterine shunt placement by a university research team of surgeons should not be available to her and her child? After all, she just recently watched the evening news description of such a successful procedure. Is it any surprise then, when she presents her personal physician with this information in the form of a request and her physician is truthful with her, that this lady might be upset enough with him to sue? That being the case, would it be at all surprising that such an emotionally charged case reaching jury trial might result in a phenomenal sum of money being awarded the mother, causing malpractice premiums and, consequently, the cost of care to rise again? Media and litigation feed off one another and are underestimated but extremely powerful players in the soaring cost of healthcare delivery. They instill medical misperceptions to millions, just as some Detroit dermatologist did for my little old Sarah and her superficial precancerous skin lesion of the nose.

If anything has kept pace with technology's growth and provided it an additional market to boot, it has been litigation. Beginning my practice in 1974, the cost of my malpractice premium was in the neighborhood of $800 and could be obtained from any major insurance company. Today, 30 years later, it costs nearer to $8,000, if you can find a company that still is willing to write it! Besides peace of mind, the other thing that malpractice insurance represents to a physician is overhead, and all shoppers know all too well who winds up paying overhead expenses. Malpractice premium cost is an area where the surgeons and procedure performing physicians might very legitimately defend some degree of higher reimbursement over their colleagues, since their malpractice premiums are much greater. Certainly it would be a token of fairness to physicians and a real eye-opener for the consuming public, if when the news media are airing one of their episodically recurring *incomes of doctors'* news specials, that next to the average income per medical specialty they show also the average malpractice premium paid by that specialty. Before the medical profession died it was not necessary to even carry malpractice insurance, and this represented a significant overhead expenditure that did not have to be passed on to the patient. Physicians then, choosing not to carry malpractice insurance, were referred to as "bare" physicians. Don't bother looking for any "bare" physicians today in the hopes that less overhead will translate to lower office charges, as "bare" physicians are nearly extinct. These brave souls are going the way of the passenger pigeon due to increasing bureaucratic and regulatory encroachment into the profession and through special interest pressures by the legal profession. The paranoid paper chase of today's hospitals has led, not only to malpractice coverage being mandatory for physicians wishing to have hospital privileges, but to hospitals actually dictating the <u>minimal coverage acceptable</u>!

I well remember in the late-seventies when my own first malpractice premium suddenly doubled, and it probably is much more than coincidence that two of the most frivolous malpractice awards I have ever heard of to date received considerable media coverage during that interval of time. One involved a several hundred thousand dollars settlement to a psychic who claimed loss of his mental powers after undergoing a CT scan of the brain. The other involved a lady who was awarded an equally ludicrous sum for pain and mental anguish allegedly suffered as a result of cosmetic surgery on her belly button leaving her navel looking, in her words, "worse than before." Press coverage came complete with before and after pictures and, although the photo quality was not great, I recall thinking at the time that a more reasonable complaint in her suit might have been that no surgery was done at all, since I really could tell no major difference one way or the other. These were the first frivolous suits that I remember, but once the door was opened, many more occurred and continue occurring today.

Litigation's hidden contribution to the medical economics equation today is eye-popping enough when considered from the perspective of the contribution of malpractice premiums to the overheads of physicians and hospitals. Eye-popping maybe, but malpractice premiums' contribution to overhead cost is minuscule compared with the effect of its greedy spawn, defensive medicine!

The term defensive medicine was coined for referring to medical decision making based not on what is actually needed for reasonable and effective case management but on what the physician or others (it's the others today becoming increasingly more important) perceive as needed for the soundest paper trail for defending against a law suite or for keeping the Joint Commission, the Centers for Medicare & Medicaid Services, or licensure happy. Simply put, the term refers to all tests, procedures, and referrals obtained beyond those things actually needed by a physician for reasonably and effectively managing a case. All within the health industry agree it's a major problem, but estimates vary regarding how major. I've heard medical opinions suggesting that anywhere from twenty-five to as high as thirty-five percent of the total cost of medicine today is the result of defensive medicine practices. My own opinion is that it is much nearer the higher figure than the lower, and that there are a number of factors on the table for assuring that it will go even higher in the future. Virtually all physicians have come to engage in this cost-rocketing practice, sometimes not realizing it, but, more and more frequently today, physicians are engaging in it under duress from the *others*. With the increasing paranoia concerning the exhaustive paper trail necessitated by the Joint Commission, hospitals have irrationally become the greatest contributor to the problem by developing a myriad of in-house committees for monitoring every aspect of physicians' actions and documenting the findings of these committees within the paper trail! In the process further paranoia is transferred to physicians causing them, not only to dot their "i"s and cross their "t"s, but to dot and cross half the remaining alphabet as well. This invariably translates into more needless testing and procedures being ordered by physicians. All such hospital paper-trail-generating committees operate and justify their existences under the guise of quality assurance, but if their origins are traced back far enough, inevitably they will lead to some mandate by federal or state government or to some special interest group's addition to the paper trail demands of the Joint Commission. Those originating from federal or state mandates are at least motivated by good intentions, but, due to bad advice given their authors and the latter's' naiveté in medical matters, they fail to accomplish any real protection against bad medical practices or achieve any savings of the Medicare dollar. Those having their origins in the annually changing requirements of the Joint Commission's paper trail, however, with each passing year, appear to be of less altruistic motivation. A proper accounting of the numbers of incomes now dependent on the Joint Commission's continued existence makes this somewhat more understandable. The larger and more complex its paper trail, the easier it is for the organization to justify their existence and perpetuate their dupe of third party carriers and hospitals that they actually provide a meaningful service with regard to quality assurance and safety. Each additional layer of the paper trail is met by increased defensive posturing by clinicians and hospital administrations in order to feel comfortable, and so, the practice of defensive medicine increases annually.

In its simplest clinical form as it applies to a physician, an example of defensive medicine would be a mother bringing her seven year old son into the doctor's office, concerned since,

"Doctor, little John has been acting puny for a week or two, and he has an eight year old cousin in St. Louis that was just diagnosed with leukemia last month and he was acting the same way when they took him to his doctor and learned he had leukemia."

The doctor listens patiently, carries out a history and a physical and finds nothing to make him suspect any significant illness, especially not leukemia. The doctor then reassures mom to this effect, or does he? Mom looks doctor directly in his eyes and with a menacing undertone to her voice asks,

"How can you be absolutely sure? All you did was ask some questions and look at him. His cousin acted just like this, and when they did a blood test he turned out to have leukemia."

In the unbridled litigious atmosphere of the day, it's time for the doctor to exchange his white coat for a white flag and order up a blood count for little John. In past years when medicine was a profession, part of which was art, the physician might have tried further at protecting little John from an unnecessary and painful venapuncture and his insurance carrier from the unnecessary additional laboratory fee by reasoning some more with the boy's mother. In today's paranoid and competitive medical workplace, it is simply quicker and perceived as safer, for the doctor to accept his cue from mom and order up the test. Even if the doctor's comfort level was sufficient for ignoring the legal repercussion tone in mom's voice, he knows that, with the number of health providers today and the resulting competition for patients who can pay, she can leave his office and take little John to someone who will do the test to please her, and he will lose a patient in the process. Impacted by a legion of such anxiety provoking outside forces, the MD's goal today is no longer an accurate diagnosis leading to a well patient. Instead, today's doctor's goal is a totally documented diagnosis supported by ancillary tests and imaging sufficient in number and quality for reproduction for family, attorney, regulatory agencies, and hospitals. For several years now most physicians have been permitting the angst of today's practice arena to translate into permission for patient, attorney, hospital, nursing home, nursing staff, and patients' families to dictate an excess of additional unnecessary procedures and tests not really needed in management of the case.

In the past, when doctors had total accountability for patient care, defensive medicine was limited to physicians only. Today, though, new philosophy by regulatory agencies has swelled the ranks of those spending money in the practice of defensive medicine. The Joint Commission, the Centers for Medicare & Medicaid Services, and state regulators have tremendously upped the ante for defensive medicine costs by insisting that more and more allied health professions take an active part in case management, whether they are needed or not. The end result has been that the paper trails of dieticians, pharmacists, occupational therapists, physical therapists, and speech pathologists have become an end to their own means in the eyes of hospital and nursing home administrators seeking the perfect paper trail and the blessing of the Joint Commission, the Centers for Medicare & Medicaid Services, and state regulators. It's not unusual today to have four, or more, additional allied paramedical professionals forcing the physician to practice even more defensive medicine and ratcheting upward the cost of care for us all in the process. Because of regulatory agencies' emphasis on documentation that *multi-disciplinary case management* is being provided for patients, nursing home and hospital administrations use every opportunity for encouraging the participation of these allied medical care providers, including outright pressure on attending physicians for consulting them, regardless of whether they are needed for an improved patient outcome or will be of any benefit to the patient.

An example of how this plays out in a real-world clinical environment would be a nursing home patient who experiences choking while eating a meal. The floor nurse makes a note in the patient's chart documenting the occurrence. The "carpet nurse", dutifully dogging the paper trail requirements of regulatory agencies, spies the floor nurse's entry. "Carpet nurse" is a term I've coined referring to those nurses no longer directly involved in clinical care (their numbers in hospitals are increasing annually) whose primary job is paper trail maintenance. The term was chosen since these individuals are easily identified within any hospital or nursing home through their offices being located far from patient care areas, in the administrative carpeted area of the facility. Their offices are tell-tale as well, having no current nursing or medical texts within their bookcases but volumes of regulatory agencies' manuals, instead. Also conspicuously absent is stethoscope, blood pressure cuff, or any other evidence of clinical patient care. Important to note for consumers is that "carpet nurses" are generally included in the count when licensing and regulatory agencies are calculating a facility's required nurse-to-patient-ratio, even though they rarely are directly involved in assessments or direct care of patients. Some of the carpet nurses then, see the note regarding choking and instruct the patient's actual floor-nurse to make sure that the doctor is told of this and that he writes a progress note in the patient's chart concerning the incident. The doctor then evaluates the patient on his next trip to the nursing home and learns that the event occurred once previously, while the patient had been eating some meat about two months earlier. He learns additionally that the patient is a total-nursing-care patient who is 87-years-old and has had no cognitive abilities for the last six years. Additionally, the patient has severe emphysema, precariously compensated congestive heart failure, and contractures of fingers and knees. In short, the patient is an 87-year-old in her end-stage-of-life and devoid of any meaningful quality of life (except the pleasure of eating). The doctor orders her diet be changed to a mechanical soft diet, with all meats to be ground, and proceeds toward his next stop within the facility. The patient's floor-nurse quickly scans the chart and calls to the doctor before he has gotten off the ward,

"Doctor did you write a progress note?"

The doctor returns and dutifully places a note in the progress section.

"Patient reported to have two episodes of choking during meals over the past two months. Both occurred while the patient was eating meat. We will change diet to mechanical soft with ground meats."

Within the week the carpet nurse calls the patient's ward and asks the nurse to read her the doctor's orders and progress note. Having done so, she then instructs the floor-nurse,

"See if you can get the doctor to write an order for the dietician to consult on the case. OIG dinged us good [sic] on their last inspection because we weren't taking advantage of our dieticians, and so, we developed a policy after their last inspection saying we would involve dieticians wherever weight loss or inability to tolerate diet is suspected. The forms we use are in the back of the chart, and if nothing's on them it will look to a Medicare reviewer or to OIG like we aren't following our policies."

The doctor is called in the middle of his busy office hours with the floor nurse's message.

"Fine, have the dietician consult on the patient," the busy doc responds.

Within another week the dietician reviews the patient's chart, interviews the nursing staff, and tells the ward nurse,

"I would continue the present diet as ordered if she is not having problems with it. But my dietary consult sheet for the chart has a box to be checked under swallowing problems as to whether or not a speech therapist was consulted. I would really appreciate it if you would tell the doctor that I said a speech therapist for evaluating the patient's swallowing ability might be a good thought. Also, the record of my consultation has a section on it to be filled in for all weight-loss cases, and I see that this lady has lost seven pounds over the last three months. You might ask the doctor if he minds ordering a pre-albumin level and a hemoglobin and hematocrit. Our nutritional guidelines recommend we do these on all weight loss patients as indicators for possible malnutrition."

Next week the speech pathologist arrives and carries out a bedside swallowing evaluation. Finished, she writes on her consult sheet in the patient's chart,

"This patient did reasonably well with her bedside swallowing evaluation. I think she just needs some follow-up visits for training in swallowing techniques, and she will probably do fine."

Handing the chart back to the nurse, the speech therapist compulsively muses to herself,

"I'd sure hate for anything to happen like a major aspiration (food or water getting into the lungs during swallowing) event or death due to pneumonia from any cause, after I've put that recommendation in writing in the chart. Just to be sure, I'll order up a "cookie swallow" (an x-ray done while the patient swallows a small amount of contrast dye for looking for food getting into the trachea) down at the hospital. I'll let you know when it's scheduled so you can get the doctor to write an order for her transfer out there for the X-ray on that day."

"She'll have to go by ambulance with everything else she has going on," the floor nurse warns.

"Well, I'm sure the doctor will write that order. I can't do that", the dietician fires back. Three days later the doctor is called by the nursing home nurse again.

"Doctor, we need an order for Mrs. James to be transported by ambulance on Tuesday next week to the hospital for a cookie swallow test.

"A cookie swallow test, who wants that?"

"The speech pathologist wants it," quips the nurse. "She says she wants to be sure there are no silent aspirations going on that might lead to pneumonia."

"Ms. Jones, how many fevers or pneumonias do you remember Mrs. James having this last year?"

"Well, none, off hand."

"Exactly, and I wonder how much a radiologist's fee for interpreting the results of a cookie swallow is these days, or, for that matter, how much the speech pathologist charges for a consult and follow-up training techniques in swallowing," grumbles the now increasingly testy doc.

"I don't have the slightest, Dr. Daley. Do you want to call the speech pathologist and talk to her about it all?"

"Don't have the time or, these days, the inclination! Send her on out for the X-rays!"

Two weeks later the doctor's office manager reminds him.

"Doctor, this is the second request from the ambulance service asking you to fill out the forms explaining why it was necessary for Mrs. James, over at the nursing home, to be transported by ambulance to the hospital for that swallowing test. Without them they can't bill Medicare for their $300."

Four weeks later Mrs. James dies quietly in her sleep at the nursing home from a blissful heart attack. Insult to injury, if the order sheet of Mrs. James's chart were examined, it would appear that the doctor was actually in charge of Mrs. James's care and responsible for ordering all her needless and patient unfriendly studies.

For several years now, most physicians have let the angst of today's practice arena and the irrational demands of the Centers for Medicare & Medicaid Services, state regulators, and the Joint Commission's paper trail requirements force them into a more lenient policy of permitting hospitals, nursing homes, and a host of other allied health professionals to dictate additional clinical tests and treatment modalities that each of them feel necessary for achieving their own safety zone as participants in the paper trail. Currently multimillions of dollars are being added to the cost of care annually by this mechanism alone. From personal experience and conversations with other physicians, I am convinced that the dollar figure today spent in defensive medicine by *non physicians* rivals that being spent by physicians at the game.

Just this past week while making rounds in the nursing home, the ward LPN handed me Mrs. Green's chart and a form for requesting a physical therapy consult.

"They want you to refer Mrs. Green to the physical therapist," she said matter-of-factly. I know without asking who the *they* are. It's the facility's carpet nurses beating the paper trail again.

"And why do *they* want that done?" I ask, with emphasis on the word they. Immediately the LPN's tone became defensive with,

"It's not us, Dr. Daley. Most of the staff actually providing the care for Mrs. Green thinks she fakes falls for attention seeking, and several of the staff have seen her do so. The problem is that our policy is to notify the patient's family of any untoward event, and today her family is complaining to the higher-ups about all the falls. So, the higher-ups say we need to have good documentation showing we have looked into the situation, and they thought a physical therapist consult for evaluating her ambulatory capacity would be a good start."

When medicine was still a profession, accountability for patient care and patient outcome rested only with the physician, and a request as ludicrous as this could have been laughed off for what it was. In today's paranoid, administrative landscape, however, a doc interested in remaining gainfully employed no longer has the luxury of doing strictly that which sound medical protocol and common-sense dictates to be in the best interest of the patient. The decision as to when consultations or additional studies are indicated should remain the exclusive purview of traditionally trained MDs, and both patient and cost of care would benefit tremendously.

One of the greatest contributors to hospital overhead comes from the Centers for Medicare & Medicaid Services' hand maiden, the Joint Commission. Paradoxically, an entity, birthed by physicians and nurtured through its infancy by physicians, has today become a thorn so painful in the collective sides of medical staffs as to lead more physicians to opt out of doing hospital care with each passing year. From the altruistic efforts of a physician hoping to provide better hospitals and better patient outcomes long ago, the Joint Commission has become today the classic example of the hell at the end of the proverbial road paved with good intentions. It's much more than coincidence that its most onerous characteristics have developed since 1965, the year that Congress passed the Federal Medicare Act. Having no experience in this part of the private sector and in need of some method of qualifying hospitals meeting Medicare's standards, Congress looked toward the already extant JCAHO (JCAHO, much later, changed its name to the Joint Commission) as a model and, in their haste, elevated and legalized JCAHO to a new role as comptroller of Medicare's purse strings. JCAHO's new windfall role enabled them to justify their existence and to assure themselves of a very lucrative future. Given the extraordinary dollar amount necessary for attaining Joint Commission's (JACHO's name today) accreditation today, it is likely that most hospitals would have long since dropped this entirely voluntary and self-imposed inspection as financially unjustifiable had the old JCAHO not been the beneficiary of the 1965 Medicare Act's name dropping. Since the Medicare Act of 1965, the resourceful and now self-serving organization changed its complexion entirely. Now with federal backing, the old JCAHO rapidly metamorphosed into the traditionally bureaucratic mode, burgeoning in size, complexity, demands, and, most significantly, budget! For once economies and livelihood's became dependent on the monster, more and more of its resources began to be dedicated to servicing overhead and for justifying its continued existence. One on-line consumer organization suggests that in addition to conducting accreditation surveys and its charges for this service, in 1993, the organization now known as the Joint Commission, sold educational materials to hospitals on how to pass their inspections, generating 21.2 million dollars in sales!! Today the Joint Commission's overall contribution to the overhead of the nation's hospitals and, therefore, to the cost of health care in this nation is gargantuan and indefensible. An administrative official of a large psychiatric hospital shared with me that the actual charge for an inspection by the Joint Commission was a whopping $52,000, and this did not include the hundreds of salaried man-hours of the hospital staff necessary for concocting and maintaining the paper trail requirement of the Joint Commission throughout intervals between its inspections. The latter is the real budget buster! Such exhaustive paper trails are vital to the Commission's inspection teams so they can get in and out of the facility within their scheduled two to three days time and for supporting the illusion that some meaningful evaluation has been performed in the process. The same administrative individual informed me that were the hospital staff's manpower hours accounted for in the cost of the inspection, the cost for the survey in his estimation would have been closer to $500,000 or even higher. Anyone who has ever shopped knows that this cost is not ultimately the hospital's problem, since we all know only too well where the cost of doing business in the retail world shows up.

Aside from the astounding overhead added to a hospital's cost of doing business, does the now all-out quest for The Joint Commission Accreditation Plaque for display in the hospital's lobby affect anyone else? The answer depends on whether you mean in a positive or negative fashion. Ask any physician how much time is spent on hospital committees generating the Joint Commission's mandated paper trail, and then ask them where they make that time up during the course of their remaining clinical day. In the hospital, time spent in committees and record keeping is time stolen from the clinical practice of medicine and care of patients. This problem is even more applicable in the case of nurses since, as employees of the hospital, they must wear many different hats, at their employer's discretion. Fear of loss of the Medicare dollar has led today's hospital administrations to near hysterical paranoia at the thought of an inadequate paper trail. Indeed, hospitals' administrations' quest for the perfect paper trail has become so intense and the process so complex that it has required the more educated, gifted, and experienced of the nursing and administrative staff for comprehending and dealing with it. This brain drain has resulted in a profound negative impact on clinical nursing and, subsequently, the quality of care for patients. Any degreed nurse will attest that with each passing year more and more of his or her salaried time is spent in record keeping and documentation, with less and less at hands-on clinical care of patients.

The nursing schools themselves today, also acutely aware of the emphasis being given the paper trail by hospitals, are now presenting curricula orientated more and more toward documentation, legal issues, and paper trail generation at the expense of basic medical sciences and clinical skills training. This trend is resulting in many nurses today entering their first clinical employment with much less medical science knowledge and fewer clinical skills than their predecessors of past decades possessed. As a result, they are hired and then receive, from on the job training, the clinical knowledge which used to be taught them in their formal nursing education programs. Boding ill for the patients in all this is that with the paper trail gobbling up the time of the more educated and experienced nurses, the hands-on day to day clinical care of the patient is left by default to the newly graduated, less experienced, and less medically grounded nurses, usually newly hired nurses from the two-year degree programs. The amount of formal nursing education and clinical experience possessed by a nurse is critically important to patient care and case outcomes, as a study which appeared in the *Journal of the American Medical Association* suggested. This study reviewed data from 168 Pennsylvania hospitals and found death rates among patients undergoing common surgical procedures to be nearly twice as high in facilities where the percentage of nurses with bachelor's degrees was low. This has incredible implications for consumers today when the preponderance of newly hired nurses in most of our nation's hospitals is now from the two-year degreed programs. As a physician, if I were a patient having undergone a difficult cystoscopy in the afternoon and calling my eleven-to-seven-shift nurse that I was having a chill, it certainly would be important to me that she had clinical background and training sufficient for understanding the potential for a life threatening sepsis (infection in the bloodstream that can occur as a result of invasive instrumentation procedures) presenting in just this manner and notify my doctor at once. Today, though, my fear is that I would have a newly graduated nurse, minimally educated in clinical sciences, who might offer me an extra blanket or some Tylenol for my chill and then hurry back to her paper work since her shift was ending and she knows all too well how her performance will be judged by her supervisor should the paper trail documentation get short changed. Nursing's and hospital administrations' switch of their priority focus from clinical care of the patient to chart, policy, and records management is directly due to the complex and continuously changing requirements for information to be included within the records for the Joint Commission's viewing at their triennial hospital surveys.

Aside from the quality of care issues created by the Joint Commission's diversion of hospital resources from patient care to administrative record management, some of that organization's more creative special-focus projects also have seemed extremely ill conceived and, arguably, have resulted in profoundly negative and unintended consequences throughout the health care system.

In particular, the Joint Commission's 2001 *Pain the Fifth Vital Sign* campaign comes to mind. Apparently concluding, by their own mysterious means, that acute pain within the hospital setting was being under evaluated and under treated, the Joint Commission arbitrarily decided to place themselves in charge of all things related to pain within our nation's hospitals. As per usual their publicity for their latest wild goose chase was announced well in advance to hospital managements across the nation and was received by the majority of them with the usual unbridled hysteria specifically reserved by hospital CEOs for anything and everything having to do with the Joint Commission. Within a mercifully short time, however, the hysteria in our hospital transformed itself into the now standard response to all the Joint Commission's capricious edicts: more meetings, more new policies, more forms stuffed into the charts for nursing to wrestle with, and more daily distractions from the business of clinical care of the patients in the interim until the Joint Commission finally arrived for their inspection, some eighteen months later.

During the Joint Commission's Pain the Fifth Vital Sign campaign, I was serving as in-house medical officer for a 150-bed state psychiatric hospital and was appalled one day to find notices suddenly posted on the wards for reminding patients that they had the right to have evaluation and treatment for any pain issues they were experiencing. Additionally the nursing staff was now, at the beginning of each shift, specifically asking patients if they were experiencing any pain, and, if so, was it being adequately taken care of!! Human nature being what it is, this approach would be ill-conceived and fraught with problems even in a general medical hospital population, let alone in a psychiatric hospital where close to forty percent of the patients were dual diagnosis patients i.e., they had both a psychiatric diagnosis and a substance abuse diagnosis. This resulted in pandemonium on the wards and a nightmarish and dangerous milieu for staff to function in for several months, until common sense eventually prevailed. Two years after this nightmarish experience, I had the occasion for retracing a trip via two-lane highway through parts of rural Kentucky, Tennessee, and Georgia which I had last made in 2000, one year prior to the Joint Commission instituting its *Pain the Fifth Vital Sign* campaign. On this return trip I could not help noticing that the medical landscape for at least a third of the small communities had significantly been transformed in the short interval since my previous trip. Each of them now boasted a medical office not previously present. Sometimes these new medical practices were located in old store-fronts or sometimes within buildings that formerly were small residences, but all of them were calling themselves "Pain Centers" or "Pain Management Clinics". None of them, from their doctors' titles or from size and appearance of their buildings, gave any reason for believing that they were staffed with a board certified pain management physician, psychologist, physical therapist, and anesthesiologist as would be expected to be the case in a bona fide medical-mainstream pain management clinic. Most of them having only a single physician's name on the office shingle, it is almost certain that these "pain clinics" all were the result of a physician having been emboldened sufficiently by the Joint Commission's *Pain the Fifth Vital Sign* campaign for deciding to limit his/her practice to seeing only patients in need of pain medications. After all, if the patients were carefully selected, the standards for narcotic dosages and amounts dispensed abided by, and a meticulous paper trail kept, there was minimal risk to a physician opting to design his/her practice for this purpose. In fact, the potential for a stress free and lucrative practice without any night call was quite good! From the time of that return trip and continuing to date, the news media has been filled with stories of skyrocketing problems of misuse and abuse of prescription pain medications, physicians of all specialties having their licenses for prescribing narcotics revoked, over-doses and deaths among all age groups due to prescription narcotics at an all time high, and crimes involving prescription drugs on the rise---all since 2001, the year of the Joint Commission's *Pain the Fifth Vital Sign* campaign. In October 2009, the AARP (American Association of Retired Persons) bulletin contained a listing of the top-50 most-prescribed drugs in the U.S. The number one position on that list was held by *hydrocodone*, a narcotic pain medication having 123 million three hundred thousand prescriptions written for it in 2008, at a retail cost of around 1.78 billion dollars.

So, today we are at the beck and call of this budget and resource gulping monster called the Joint Commission, without any real evidence that patient care or outcomes benefit whatsoever from its increasingly expensive and outrageous demands. In my experience, a Joint Commission accreditation plaque in the lobby of the hospital affords as much real protection for that hospital's patients as does the membership plaque from the Better Business Bureau in a hardware store for the store's shoppers, but the price for a Joint Commission plaque is in the hundreds of thousands of dollars annually and, most importantly, the Joint Commission plaque is being much more seriously accepted by patients as a guarantee of safety and quality.

I personally know of a small community hospital that had been owned and managed for many years by an order of Catholic nuns. During those years the facility was held in high esteem by patients and physicians alike. Unfortunately, as the physical plant grew older, the nuns had increasing difficulties with each Joint Commission survey due to the aging physical plant. Finally, the Joint Commission demanded that major upgrades to the physical plant be underway by next survey or accreditation would not be forthcoming. The nuns did not have the resources for refurbishing or rebuilding. And so, they made the painful decision to sell their hospital to one of the largest hospital management corporations in the country. Soon after assuming ownership of the hospital the management corporation learned it was close to time for the next Joint Commission survey. As one physician colleague on staff there told me, "It was absolutely amazing. They (the new management) waxed some floors, painted some walls, and passed inspection without any problem whatsoever!" Even more incredible according to him, absolutely no major renovations to the physical plant were made until several years later! Clearly, the hospital management firm played the paper trail game much more cleverly than the nuns or, perhaps, more *cleverly than the nuns were willing to play it* .

Recently I was required to give up a half hour of my hospital time for attending a meeting where the medical staff was to be briefed on the new charting forms the hospital was implementing. The forms were described in detail by a well-intentioned social worker who had chaired the revision committee. At one point she was asked why a particular page was so cramped for space. A physician asking the question suggested that one of that cluttered page's categories more logically fit, for the sake of medical continuity, in another area of the chart, to which she replied,

"Oh no, Joint Commission doesn't like to spend time looking through the chart for things. They like everything right there together. It makes their job easier," the social worker replied in dead earnestness.

HELLO!!! How long is it going to take hospital administrations to realize that ninety percent of labor-intensive paper trail innovations are for making the Joint Commission's, the Centers for Medicare & Medicaid Services, and states' regulators' jobs easier— not for making them more meaningful where quality of care is concerned.

Common sense tells us that, if someone is watching, it's a thousand times more difficult to fake giving an ordered 8:00 p.m. enema when you're dead tired, behind schedule, and understaffed than it is to use paperwork for the deception. A check mark in a box on the appropriate sheet, and the enema is seen as done. Actual oversight, however, means increased overhead and decreased bottom lines, a situation not favored by the business firms managing most of our hospitals today. They prefer, instead, on spending much larger overhead dollars on an ineffective Joint Commission. Where a premium is placed on documents, the temptation for taking the easy way is overwhelming. The late Ray Crock, founder of McDonald's hamburger chain was well known for his impromptu inspection visits to his restaurants throughout the country, often in the guise of a customer. I'm sure he had reams of reports and documented inspections at his beck and call, but he knew that paper trails are vulnerable to quick fixes and frequently not representative of the facts.

I also was required to give up an hour of my hospital time for attending a training session set up by hospital management for the purpose of readying the medical staff for an expected up-coming Joint Commission inspection of the hospital.

By that hour's end I was a little nauseated and a lot angered. In that sixty minute period I heard, "Joint Commission" at least a dozen times, "good documentation" at least eight times, and "Pratt and Smithe" about five times. Sadly, I did not hear the words "patient", "quality", "care" or "outcomes" a single time! My nausea peaked with realizing that "Pratt and Smithe", referred to so frequently during the session, were the two enterprising and opportunistic individuals conducting the training session, who, having recognized the degree and extent of paranoia of hospital administrations regarding the Joint

Commission's paper trail, had parlayed it into a lucrative and thriving business as consultants for passing Joint Commission's inspections. In so doing they joined the layers of numerous other accounting, recording, and documenting businesses, all making a handsome living as leeches on the medical system and inflating the cost of healthcare for us all. This was definitely a first for me and added financial injury to insult. Now our hospital was passing on to its patients the cost of a Joint Commission pre-inspection inspection!!

The number of full-time hospital employees and contracted services whose only contributions are generating and organizing an easily accessible paper trail for Joint Commission, Centers for Medicare & Medicaid, and other regulatory agencies, would shock even a seasoned insider to the industry were it ever to be accurately tallied. A cursory glance would reveal most of the medical records department, risk management department, and quality assurance department, to name only the few whose visible major purposes are Joint Commission mandated paper trail generation. These major players, however, are merely the visible portion of the iceberg that is out on the surface for all to see. The far-larger contingency of Joint Commission spawned workhorses consists of the one or two people, per every single department within the hospital, that are dedicated daily by the hospital administration to playing the Joint Commission paper trail game continuously throughout the year. The manpower and salaried hours represented by this group is tremendous and reflects an extremely significant portion of every hospital's overhead, likely multimillions of dollars annually on a national scale.

The hospital department suffering most in this category is that of nursing. The reason for this is simple. The major players such as quality assurance and risk management are, by training, nonmedical people. They must have a liaison capable of bridging the gap of clinical terminology and medical processes. Their job security depends on successfully bridging this gap. Additionally, these major player departments are relatively high in the chain of command within the hospital and have easy access to hospital leadership's ears. Positioned thus, they are capable of exerting extreme leverage on the nursing department to perform for them. Thus, the brunt of the burden falls on nursing, with more and more of their time being occupied by documentation, policy formulations, and data management and, consequently, with less and less time spent in direct nursing care of the patients. Even the most Joint-Commission-paranoid of hospital administrations, however, would balk at paying nursing scale wages for paper pushing, especially if it meant having to hire additional nurses to care for the patients. Having already relegated their dollars to that which they felt most important i.e., the paper trail, the administrations' built-in answer for balancing the nursing budget resides with the concept of nursing teams and supervisors. In this concept the most-educated, experienced, and talented of the nursing staff are made supervisors. The next most qualified are made unit leaders, next comes team leaders, and so on. It's a given that the bulk of the supervisors' and unit leaders' time will be occupied by administrative tasks and paper trail needs, but not as widely appreciated is the fact that even the RN clinically assigned to the patient spends most of his/her time in assuring that paper trail documentation within the patient's chart is carried out. For the patient, the end result of this approach is having the least experienced nurses (increasingly more frequently, LPNs or sometimes, only patient aids) observing, evaluating, and serving as their primary link to medical care. More evident in some than others but a fact in every hospital today, the patient has taken a back seat with regard to the hospital's focus and resources. In the driver's seat for sometime now has been the quest for the ideal paper trail for appeasing the Joint Commission, the Centers for Medicare & Medicaid Services, and State Licensure and Regulation. With each passing year, administrative costs devour an increasing percentage of our nation's hospitals' annual budgets, and this cost of doing business is passed on to the patients or to their insurance carriers.

On the rare occasions you do hear Joint Commission using phrases such as "patient needs" or "quality outcome", you can bet they are in their justifying-their-existence mode for media or public. Yet, to my knowledge, nothing exists for documenting that they have positively impacted the quality of patient care or outcomes whatsoever. In fact, a behind-the-scenes investigation by CBS's *60 Minutes* of a psychiatric hospital managed by one of the nation's largest private psychiatric hospital management corporations, clearly proved this to millions of TV viewers. Using an on-site secretly wired and videoed social worker posing as a new employee at the hospital, they unearthed a litany of extremely serious problems including: understaffing, lack of training, lack of supervision and very questionable patient care. It was blatantly obvious to even a lay person watching through this planted social worker's hidden audio and video, that not only were some patients in the facility receiving virtually no meaningful care or services, many were there against their will and, apparently, with little justification. The scenes within that facility and the candid conversations among the staff were absolutely chilling to any responsible healthcare professional and clearly suggested to anyone of even common-sense that the name of the game at that particular facility appeared to be revenue, not treatment. All this was revealed by actually being physically present as an unexpected observer, not by relying on a rehearsed and well-prepared paper trail for a description of how the hospital allegedly functioned. This point was masterfully brought out near the end of the telecast when news caster Ed Bradley took the most pertinent question of all directly to a Joint Commission representative responsible for evaluating that facility. Pointing out that this facility had recently been inspected and awarded Joint Commission's accreditation only a short time prior to this undercover investigation, Mr. Bradley essentially challenged the Joint Commission representative to explain to him how a hospital, having such questionable patient care practices as had just been witnessed by a national TV audience, was able to receive Joint Commission's blessing and accreditation. The reply to the veteran news correspondent's question, by the now totally flustered Joint Commission representative, was delivered with a chagrinned tone and essentially consisted of an admission that the brunt of his organization's inspection consisted of records and policies review. Moreover, he went on to admit that, without someone telling his team of what transpired within the facility on a daily basis, the Joint Commission would have no way of knowing!

With that simple but all too accurate admission by the Joint Commission representative, this author rests his case. Evaluating the paper trail of hospitals' policies and records doesn't give a clue as to what really occurs to the patients within them.

Presently the vehicle that hospitals are depending upon for delivering their Medicare dollars is a broken, out of control, gas guzzling, bankrupting, and logistical nightmare driven by special interests instead of the patients it was purchased for. The Centers for Medicare & Medicaid Services is the federal agency charged with the responsibility of disbursing funds to hospitals for services rendered to Medicare patients. As such it is counter intuitive that an independent organization (the Joint Commission), which the Centers for Medicare & Medicaid Services has no control or oversight of, retains absolute authority for designating which hospitals are deemed of suitable quality and service for receiving the Centers for Medicare & Medicaid

Services' money. Due to the wording of the 1965 Medicare Act, however, this is precisely the case and it desperately needs to be corrected. As a result of a 2004 U.S. Government Accountability Office report, *Medicare: CMS Needs Additional Authority to Adequately Oversee Patient Safety in Hospitals*, it should not be news to many in the halls of Congress that the Joint Commission has been ineffective at assuring quality and safety within our nation's hospitals. This report alleges that the pre-2004 Joint Commission hospital accreditation process failed to identify many hospitals that follow-up inspections by other agencies identified as having Medicare program deficiencies. In fact, it appears from the report that in a sampling of 500 Joint Commission accredited hospitals, validation follow-up surveys conducted in 2000–2002 revealed 157 of these hospitals to have Medicare deficiencies, and that Joint Commission failed to identify 123 out of the 157 i.e., 78 percent had been missed!! By eliminating the ineffective and exorbitantly expensive Joint Commission way of doing business and replacing it with a meaningful, practical, and economical process, Congress could make a giant step towards the *real change* in the health care system that they profess is their goal. To effect such a change would require Congress to rewrite or amend the 1965 Medicare Act, a task which they might have little appetite for. Their appetites might be stimulated, however, by their asking both the GAO (Government Accountability Office) and CBO (Congressional Budget Office) for a detailed study on the year-round hidden cost to the nation's hospitals currently inherent for participation in JCAHO's ineffective accreditation process and how this cost contributes to the cost of health care. By amending the 1965 Medicare Act and replacing Joint Commission with a meaningful and effective hospital survey process, Congress could immediately improve quality and safety in the nation's hospitals and, with proper monitoring for assuring that the hospital's resultant savings under the new inspection process go towards decreasing the fees charged patients for hospital services, health care costs could be driven down dramatically.

CHAPTER 5

Along with personal service and pride in the workplace, the medical profession can never be resurrected. We are stuck with medicine, the business. Before being accused of *if-not-part-of-the-solution, then-part-of-the-problem*, experiences over three decades as an inside player privy to the problems within our health care system have provided me with some ideas for at least transforming *medicine the business* into a more efficient, economical, quality, and patient-focused one than it currently is.

Firstly, the public and the media must clearly understand that in the eighties and nineties there were more than enough physicians, likely too many in areas of the country having highly sought after living conditions or geographies. There may have been a misdistribution, but certainly there was no dire shortage of doctors as the public was continuously being misinformed by media and special interests. Today, there likely are still sufficient numbers of physicians overall, but too many of them are earning their living in the high-pay, non primary-care specialties. Even with the predicted increased health needs projected as a result of aging baby boomers, it is likely that the current gearing up of our medical schools will overshoot the actual numbers needed and almost a certainty, the types of specialties not needed.

Randomly increasing the numbers of physicians in *all* the specialties is an economically dangerous thing to be doing given that Medicare coffers are nearly empty and some recent studies strongly suggesting that the more Medicare dollars being spent per patient in a given state, the lower that state's health care quality ranking tended to be. Other studies of medical utilization patterns have shown that in geographic regions where patient health care expenditures were high, patients received *sixty percent more* medical services than in areas of the country where expenditures were low. In the regions of high medical expenditure patients received more frequent tests, procedures, specialty consultations, and more frequent admissions to hospitals, simply because the service was available---yet fared *no better* in their overall outcomes or in personal satisfaction with their care than patients in regions who accessed care less and spent much less. In fact, the high spender group seemed to do worse, reflecting, perhaps, what all adequately educated physicians already know but are increasingly not heeding— that no medical test, procedure, or hospital stay is without risk of adverse events and complications. In other words, it appears that much of the high cost of our health care delivery in the USA, compared to similar sized industrialized nations, is due to over-utilization, especially the indiscriminate overuse of terribly expensive high-tech imaging and other expensive diagnostic studies.

Due to the media's fascination with medical technology and the omni present and hyped medical TV dramas, the public today wants and expects high-tech testing to be involved when they visit their MD with a symptom they are worried about. MDs will argue, because the public expects infallibility and wants the test, they go along with the patient and order it, chalking it up as another instance of the need to practice medicine defensively. Defensive medicine is undoubtedly a huge problem and represents a very significant reason why physicians are over-utilizing testing and imaging across the health care system, but as a practicing physician for over three decades, it has been my observation that equally frequently the expensive high-tech testing is being ordered for self-serving reasons of the health care providers.

Take, for example, a patient visiting the doctor because of headaches he has been experiencing more often than usual. A neurologist or an MD in primary care can perform a thorough history and physical exam and, with a high and acceptable degree of accuracy, determine the headache's cause. However, in the time required for carrying out such a history and physical, the doctor could see two or three additional patients, and he receives too nearly the same office visit fee regardless the method used for arriving at the diagnosis. Even if he uses the proper billing code for indicating a complete history and physical exam were done, the amount awarded him for the time spent does not equal being able to bill for an additional three patients he might work in for the same amount of time expended performing the complete history and physical on the one headache patient. Also, given that the patient comes in to the doctor already expecting an MRI of the head as he saw ordered by the doctor on the TV medical drama last week and where (surprise, surprise) the patient's headache turned out to be due to an extremely rare and malignant brain tumor, the physician now would have to spend an additional half-hour in discussing, educating, and convincing the patient why no testing was needed in his particular instance. No, it simply is more expedient and profitable for the doctor to order the testing and move on to the next patient. Yet, the ability to perform a proper history and physical exam and to process the findings through an extensive pathological and physiological knowledge base obtained through years of medical school education, is what enables a physician to be highly accurate in arriving at the proper diagnosis and to only need expensive ancillary testing for confirming the diagnosis in a limited and select number of instances. This is precisely the reason physicians are paid high fees and the justification for it. If legal authority for ordering high-tech, expensive diagnostic tests were all it took to be an efficient, safe, and effective doctor, we could reduce medical school time to six months and soon have doctors' offices on every street corner, but the quality of care and safety for the patient would be totally unacceptable and the economics totally unsustainable. This is not that difficult to understand. Consider the hypothetical situation of giant retailer Wal-Mart deciding to make available MRI head scans for any of their customers worried that the headaches they are having are due to a brain tumor. Left up to the customer, whether to have a scan or not, the number of scans performed before one brain tumor was actually found would be so high that the cost per brain tumor diagnosed in this manner would be astronomical, perhaps hundreds of thousands of dollars per each tumor found. Of course, the amount spent by customers on scans whose headaches turned out being due to migraine headaches or muscle contraction headaches would also be astronomical and unnecessary as well.

Unfortunately, the reality is that our system today is already half-way there to becoming the over-doctored and medically under-educated one predicted as a consequence from a hypothetical reduction in medical school education time from four years to six months, as previously suggested. We are likely way more than half-way there in the case of primary care physicians. This has come about due to the few remaining physicians in primary care today relying on imaging and testing for arriving at a diagnosis rather than utilizing their extensive educations and clinical skills and, increasingly, thru primary care nurse practitioners being utilized as first-contact care providers with unrestricted privileges for ordering expensive imaging and testing. This latter group, lacking sufficient medical sciences education for safely and effectively arriving at a diagnosis by any other means, tends to rely heavily on testing and imaging. Lacking anything near the medical knowledge and acumen of MDs, in the interest of safety, it probably is a good thing for their patients that primary care nurse practitioners error on the side of caution by ordering more imaging, testing, and specialty consultations in evaluating patients. It definitely has not, however, been a good thing for the economics of our health care system. Aside from potential safety and quality issues associated with increased utilization of primary care nurse practitioners as first contact medical providers, a more practical economic point simply is, at a time when our health care system's cost is out of control and Medicare approaching bankruptcy, this is not the time to be diminishing the experience and educational requirements necessary for the privilege of ordering expensive tests and imaging procedures. Instead, we should be more aggressively monitoring for the wide-spread physician inappropriate utilization of testing and imaging and demanding that all doctors begin using the extensive medical sciences education and clinical training they invested ten or more years in obtaining and return to ordering testing and imaging in the appropriate and responsible manner they were taught in medical school.

All physicians were taught in medical school that testing should be ordered only for confirming a condition or diagnosis— *that the physician's clinical history and exam has already led him to be reasonably certain exists*! Use of testing in any other manner is simply *screening* for the condition. Unlike diagnosing, screening for a condition is always expensive and can result in false-positive findings and unexpected abnormalities of undetermined significance that then need to be explained to the patient's satisfaction and, thus, leads to further testing for doing so. Because of this, medical science protocols have very strict qualifications defining illnesses for which it is appropriate and cost efficient to screen for. The protocols are based on in-depth statistical analysis and are in place to prevent the patient from undergoing expensive tests that are not likely to find anything and to prevent the finding of false-positive results which leads to further unnecessary testing, expense, and the always-present possibility of injury or illness caused by the test itself. Today, unfortunately, clinical diagnosing by physicians has taken a back seat to expensive high-tech *screening* throughout the health care system. Over-utilization of sophisticated and expensive testing has nearly become the norm within most clinics, hospitals, and ERs throughout the country. The venue prompting physicians to order needless and expensive tests for convenience and expedience more often than all others, however, is in nursing homes through out the nation.

By definition nursing home patients will experience more new complaints and more frequent declines from their baseline health status on any given day than any other patient population. Frailty and instability in multiple organ systems of their aged bodies is precisely the reason they are no longer able to live independently in their own homes or in the homes of their families. Paradoxically, the facilities we then place them in for addressing their daily high-maintenance health care needs are the facilities having the least physician-presence of all other health care venues within the medical system. Doctors' offices, clinics, and hospitals have on-site physician availability several hours out of the day on most days of the week but, not so, the majority of nursing homes. Consequently, most patient interaction, patient assessments, and medical interventions within our nation's nursing home facilities are carried out by the nursing staff, and all too frequently the ratio of licensed nursing staff (LPNs and RNs) to patient aids on these staffs is the minimum permissible by federal and state regulatory agencies. Making matters worse is that the bulk of the time of the licensed nursing staff is taken up with charting, record keeping, and supervisory tasks, leaving them little time for interacting with and clinically assessing individual patients. This results in a resident's complaint or changes in status usually having to be brought to the licensed staff's attention by a patient aid recognizing that something is wrong with the patient. The aid then has to go to the nurse's station and interrupt a licensed nursing staff from her paperwork duties and relay to this individual her observation or concerns.

A typical patient medical intervention occurring within the nursing home often goes like this. The only licensed nursing staff for the unit is sequestered within the nursing station, intent on fulfilling the one aspect of his/ her job description they know all too well will be scrutinized by their superiors and which can make or break their job security. In this typical illustrative example, let us assume the nurse is a young lady recently graduated from a 2-year associate RN degree program. She is acutely aware that her creation of an adequate paper trail, should facility administration or, worse yet, a regulatory agency ask for it, is what her superiors consider to be her priority nursing responsibility. Now suddenly jolted from concentration on her charting by the barely disguised fear detectable in the voice of the patient aid delivering the message that Mrs. Jones in 212 "does not look good", the licensed staff member feels the first pangs of fear within her own stomach. After all, she has only worked here for five months, and this was her first job after graduating from the local community college's 2-year associate RN program. She could have interviewed for a position in the outpatient surgery center or answered the ad for a position at the community general hospital, but there had been many more positions available among the area's nursing homes to choose from. She knows, however, that a large part of her reason for choosing the nursing home, over trying for the surgery center or hospital positions, was that she really hadn't felt comfortable about her clinical skills or clinical judgment being adequate for assuming the responsibility for acutely and severely ill or injured patients. After all, two years was not much time for suddenly becoming a nurse, and it seemed to her that the bulk of that short time had been focused on nursing theory, care plan formulation, legalities, and documentation responsibilities. Oh, she could take a patient's vital signs if no LPN or patient aid was available for doing it, but, as to knowing anything about what vital signs at either extreme meant, well, that was another matter. She still wasn't sure she could distinguish a regularly irregular pulse from an irregularly irregular one and had no idea of what significance either finding might hold for a patient having them. She had admitted this to one of her classmates once and had been advised "not to worry" since it wouldn't really matter as she would just be calling the doctor and telling him/her about it. Her classmate's response had sounded reasonable at the time, but so far in her first real job as a nurse here in the nursing home, she had learned that calling a doctor was easy enough but getting one to return her call promptly was not. Even when she reached a doctor by phone it was usually never a positive experience for her. They always seemed in a rush to get off the phone and back to their offices full of patients or to get back to making their hospital rounds, depending on where she had reached them with her call. It often seemed to her that she and her nursing home patients were considered by most of the doctors to be of very low priority, almost as if they were volunteer work that the doctor would get around to at his convenience after his office and hospital patients had been attended to. She knew, of course, that this was not the case, and the doctor was being paid for seeing her patients, usually by Medicare or Medicaid in most instances. All the same, in the majority of instances where she had conversed with a doctor regarding one of her patients it had been by phone, and it seemed evident to her that the goal of the physician at the other end of the line was to address her concerns in the shortest phone time possible and with the least inconvenience and risk to him. Thus it usually was the case that, if the patient's condition was reasonably stable, an order would be given by the doctor for addressing the symptom sign, or situation of the patient as described for him by the nurse, followed usually by the admonition to call him/her back if that didn't take care of the problem. At least half of the time the doctor's initial orders did not take care of the problem, thus requiring repeated phone tag between herself and the doctor. It differed from doctor to doctor as to the number of additional calls and additional phone orders trialed before the order, "ship to the emergency room" would come, but come it always would. What never came, however, was the physician to the nursing home to actually evaluate the patient and write orders for treatment along with a personal explanatory progress note. In the nurse's mind she knew that the doctor coming on-site and evaluating the patient would have gone a long way toward making her feel that she had someone sharing the responsibility of the patient's outcome with her; someone with more knowledge, education, and, most important of all to her, authority. In a nut shell, the doctor's coming and evaluating the patient would have alleviated her very real concern that the patient might expire there in the nursing home with herself being the only and highest licensed medical staff present. Now entering Mrs. Jones's room and observing her tiny frail body with eyes closed and chest rapidly raising up and down beneath her gown, limited though her experience was, the young nurse is not surprised when Mrs. Jones does not open her eyes or even stir at the sound of her name being called directly into her little blue-tinged ear. The nurse thinks to herself that the patient aid had been right. This well could be the night that this poor soul would finally find relief from her suffering of the past year or more. Her thoughts quickly jump ahead to calling the doctor and to what he will likely order, and she suddenly realizes that it isn't this poor eighty-six year-old lady's pending death that makes her so uncomfortable. After all, the entire staff has been expecting this poor soul's death for the past two years, many of them silently hoping for her sake that it would come peacefully during her sleep and soon. No, it is not death that makes her anxious. It's her feeling of vulnerability to the system at being the only medical representative present should administration, regulatory surveyors or family members question, after the fact, if something more could or should have been done for the dying patient. For her own part, even this early in her career, she is quite certain in her own mind and heart that no medical intervention with the exception of good nursing care, dignity measures and medication for comfort will be of any benefit to this patient. In fact, she had reflected on several occasions over the past few months, while attending to the patient's catheter and gastric feeding tube care, if it were herself in this condition, she would consider death a blessing.

Also, last month when Mrs. Jones had moaned with discomfort as the ambulance attendants lifted her onto the gurney for the trip to the outpatient surgery center for placement of the gastric feeding tube, she recalled feeling sad for the patient and angry at the doctor for having ordered it. She understood what influenced the doctor's decision, but it still had angered her. She well knew why the doctor had decided to order the gastric tube as she herself had nearly been driven crazy in the month leading up to the feeding tube's placement by the constant badgering from the dietician, the director of nursing and even the nursing home's administrator. All of them had been harping about the weight loss documented since the patient had quit eating, and they had continually obsessed over what would happen to the facility should licensure or Medicare review this chart and decide that "enough" hadn't been done. She had known, as soon as the doctor had asked that she call Mrs. Jones's family guardian and explain to her that the patient was no longer able to eat and losing weight, what the doctor's feeding tube decision would be. After all, in the five months that she had been employed here she had already made several long distance calls to Mrs. Jones's guardian-niece and knew her to be emotionally labile, uncomfortable with her guardian role, medically naïve, and, worst of all, guilty that she had only visited once since her aunt had been placed in the nursing home. The nurse well knew what the outcome of another discussion with the guardian would be. It would be the same as the one last month when she had to call for informing her that Mrs. Jones had another pneumonia. Since this was her third pneumonia in the past four months, she had tried explaining to the guardian-niece that her aunt had reached the point chronologically and medically that she was in the last days to months of her physiological life; no medical intervention, regardless of its nature, could reverse that course and prevent the inevitable. Also, since Mrs. Jones had, in the two months preceding that episode, been transferred to the acute care general hospital twice before, subsequently spending several days in the ICU on each occasion and with the second of these requiring three days on a respirator and a great deal of difficulty in weaning her from it, the nurse reminded the guardian that the nursing home also was licensed as a skilled care facility and, as such, could treat Mrs. Jones's current pneumonia with IV antibiotics, oxygen delivery, nebulizer breathing treatments and medications for her comfort. She also had reminded the guardian that, by remaining in the nursing home for her treatment, the patient would be surrounded by staff that were familiar with her and looked at Mrs. Jones as one of their own family. She would like also to have added that from, the patient's point of view, the nursing home was now her home, and its familiar staff was a source of comfort for her. The nurse also considered telling Mrs. Jones's guardian niece that transferring her aunt to an unfamiliar environment amid unfamiliar faces, for what very likely could be her last hours or days, would only add to the patient's suffering, anxiety and discomfort. At the last second, however, she caught herself and refrained from sharing with the guardian this last bit of pertinent reasoning since it was no longer applicable. Three years ago it would have been entirely truthful and applicable, but the reality for the past two years was that Mrs. Jones has been a total-nursing-care patient, having physiological life but no meaningful, quality life whatsoever. For over two years now she had shown neither signs of meaningful awareness, nor had she exhibited any meaningful interaction with her environment or the staff. The guardian's response to the nurse's information is quick and as expected. Despite this, the nurse gives it one more try, reminding the guardian that during Mrs. Jones's last prolonged stay at the hospital, the attending physician there had called and asked the guardian's permission for making the patient DNR status. The guardian's reply to this reminder was quick and also as the nurse had expected.

The nurse begins hoping that the doctor will return her call quickly so that she can tell him the results of her conversation with the guardian and, especially, of the guardian's wish to send the patient back out to the hospital because, "she would just feel better with her aunt in a hospital if she might be going to die." Since there now is only an hour remaining in her shift and her charting and paper work remain undone, the nurse secretly considers it better for herself that Mrs. Jones will likely be transferred out to the hospital when the doctor returns her call since, even with the additional paperwork required by a transfer, she still would be left more time for completing her charting responsibilities, even with having to make the requisite transfer telephone calls, than would be left her after taking a bunch of phone orders from the doctor and initiating the treatments should the patient remain in-house for her treatment. So sure is she that the doctor will give the transfer order that she is tempted to go ahead and call the ambulance while awaiting his return call. After all, it's 2:00 AM and she is certain the doctor won't wish to be disturbed a second or third time from his/her sleep as well might be the case should orders be given for Mrs. Jones's treatment to be administered there. Even in the extremely unlikely event that the doctor did order treatment and comfort measures to be carried out in the nursing home, the nurse knows the doctor would change it to another admission to the general hospital within a split second of being reminded of the nature of the patient's guardian's attitude, especially the part where the guardian expressly requests that the patient be transferred to the hospital again "if she might be dying." It's now been ten minutes and the doctor has not responded to the nurse's call to his answering service. The nurse smiles to herself as she proceeds with calling the ambulance for the transfer, one more time, to the acute care hospital. Scenarios like this one are being played out daily throughout our nation's nursing homes, representing poor quality of care and multimillions of dollars in unnecessary health care costs.

In any given twenty-four hours, it is likely that thousands of unwarranted medical interventions, costing hundreds of thousands of dollars, are being ordered within our nursing homes' populations for the sake of convenience and a host of other self-serving agendas having nothing whatsoever to do with the patients' needs or best interests.

Qualitatively and economically, the nursing home venue is ideal for beginning the Herculean task of wringing out the waste and mismanagement within our health care delivery system. The Devil, of course, is always in the details, but any change would necessarily have to begin with a slow and on-going education of the general public and all employees within the nursing home industry, especially those in administrative roles. Areas of educational focus might include: 1. What CPR can and cannot do 2. Who might benefit from CPR and who never does 3. Which patients benefit from feeding tubes and which do not 4. Common myths regarding what feeding tubes can and cannot do for the patient 5. Complications and social sequelae associated with feeding tubes 6. The high premium that geriatric patients place on maintaining their ability to pursue pleasure thru traditional oral dining 7. The typical geriatric characteristic of preferring quality life over prolonged life 8. The importance of formulating an advanced directive that clearly delineates the circumstances under which and the specific medical interventions not desired 9. Importance of periodically reminding health care providers, family members and friends of the advanced directive's existence and the expectation that it will be precisely followed 10. Acute Medical conditions common to the nursing home population that should not require transfer to a general hospital for adequate treatment (which would be the majority of all acute medical needs where the nursing home is designated as a skilled care facility)

Forewarned is forearmed, and a proactive approach by nursing homes that would require the families and, specifically, the legal guardians of all potential new residents to attend a mini course, covering the same ten topics as previously suggested as required training for nursing home staff, would be quite helpful and, hopefully, preventative. Ideally, the physician caring for the new resident would be present at this session for sharing his/her view on these topics as well as for answering any other questions that the new patients or their guardians might have.

The third component for addressing the waste and mismanaged care in nursing homes is an efficient mechanism for monitoring for health care providers who persistently order patient interventions based on convenience, risk management and other self-serving agendas, along with an effective mechanism for reining these individuals in. It is reasonable to believe that if the needless, repetitive ping-ponging of nursing home patients back and forth between acute care hospitals emergency rooms, and back to nursing homes repeatedly in their last days to months of life could be successfully stopped, Medicare's financial footing would become firm again and remain so well into the foreseeable future. The key for accomplishing this is bringing about a paradigm shift for the general public regarding what constitutes quality care for geriatric patients in general and in the end-of-life setting, specifically. Geriatric patients still of sound mind will testify that they don't define quality care in the same manner as do other patient populations, and, within their age-group, quality time is generally valued over quantity time, more is frequently not better, and needless suffering and futile interventions are always bad. Educating the general public to these facts will need to be done slowly, repeatedly, consistently, and the process will require a sustained joint effort from the medical sector, the media and the government. It will require a media blitz which should be on-going for years.

A much more realistic worry for the bean counters is how the proposed increase in numbers of physicians, currently being advocated for the express purpose of meeting the baby boomer crises within our nursing homes, can be remunerated, given that they are being brought on board for servicing an expanding Medicare covered population, and Medicare's funds are currently strained, with its future in grave jeopardy! Like the seventies and eighties, we likely still have sufficient numbers of physicians within the health care system today for adequately meeting the nation's needs. Most certainly we do if we are speaking of *valid* needs and commit to eliminating *perceived* needs. What we do not have, however, is enough of these physicians trained for and engaged in the primary care specialties.

Whenever even mentioning the possibility that physician supply is adequate I'm always countered with, "Well, every time I go to the doctor I have to sit and wait in an office full of patients." So what! Every time I go on a weekend to eat out I have to sit and wait forever for a table! Yet, the board managing the restaurant chain would never be naïve enough for interpreting my having to wait as justification that another of their restaurants was needed just around the corner. They know their demographics with regard to market demand, something our medical schools have completely dropped the ball on, particularly so regarding the highly-remunerated, procedural subspecialties.

Our medical schools continue to produce anatomically focused specialties such as interventional cardiologists, pulmonologists, hematologists and gastroenterologists at a constant rate, without regard for the statistical prevalence of illnesses requiring such sub specialists' attentions. Simply put, if a gastroenterologist locates in a community of 3,000 people, disease prevalence statistics dictate there will be a finite number of gastrointestinal illnesses within that community requiring the use of a colonoscope. Should that number be thirty per year, for example, the gastroenterologist would only get to perform thirty colonoscopies per year if he limits himself to only those cases with bona fide reasons for having the procedure done. This is a big *IF* for consumers wishing to avoid an unnecessary procedure. After years of training and years of borrowing educational money, thirty colonoscopies per year likely would be an unacceptable caseload for a young and ambitious gastroenterologist. His options then are to relocate, remain and gradually lose his specialty skills from lack of use or relax his indication standards for performing the procedure. The latter may occur deliberately or subconsciously, but it will occur if he stays in that community. The end result is needless colonoscopies performed just because of the availability of the service. Recent studies have made news print confirming that in regions where medical services are dense, the cost of care and subsequent total dollars spent by end of life are significantly greater than in areas with less medical service sophistication— *and with no overall difference in quality of care*!!! Twenty years ago a gastroenterologist would never have found himself in a community where there were insufficient numbers of bona fide colonoscopies to be performed and, so, be forced to perform some of questionable need for financial survival. First, there wasn't an abundance of gastroenterologists and so there was still a need somewhere, if not that particular community. Even more importantly, a *medical profession* still existed then, and the gastroenterologist's primary care community colleagues, being aware of their patients referred to him receiving colonoscopies of questionable need, would have quit referring him patients and would have complained to the hospital administration as well. Today though, there is no self-policing medical profession, and even if there were, unlike in the past when primary care physicians efficiently and competently treated eighty percent of the patients' problems and served as gate keepers to the procedural specialists, today patients make their own referrals to specialty consultation whether warranted or not, and very frequently they are not. Also, today's corporation-managed hospital administration would be much less likely to come down hard on a physician performing non indicated colonoscopies, since more colonoscopies mean more hospital revenue, and most hospitals today are run by big business corporations. Today it would likely take extremely negative media coverage concerning unnecessary colonoscopies being performed before the hospital would be all that concerned with looking into the matter.

Until our medical schools meet their responsibility for valid demographic need-studies based on disease prevalence statistics and adjust their medical schools' class sizes and specialty residency programs sizes accordingly, tremendous sums of healthcare dollars will be spent needlessly in this manner. Remember the interview with the candid young medical student of the nineties and his admission of planning to enter a specialty with a procedure associated with it, "because that's where the money is."

Along these same lines we need an objective hard scrutiny of the ongoing public and political brainwashing that such a dire primary care physician shortage exists as to justify using so-called "physician extenders", a group having only a miniscule fraction of the in-depth medical education of that possessed by an MD, as first contact primary care providers. Let alone the safety and quality issues that this practice presents for the medically naïve consumer, it adds additional fingers for dipping into the healthcare delivery payment cookie jar, and these fingers are very inefficient fingers to boot. Remember the example of the inverse relationship between depth of one's medical education and number of ancillary tests ordered as demonstrated by medical students in training. A junior medical student on his first surgery rotation will order three to four times the number of tests for being certain of the diagnosis of appendicitis than a resident in the second year of his family practice residency would order.

Recently, making matters worse, the salaries of physician extenders today have escalated to the point of approaching those of primary care physicians, a fact not lost on the nursing profession or the nursing schools operating training and credentialing programs for physician assistants and ARNPs. The managed care business loves the concept as they can continue hiring, for the time being, two of these individuals for what they would have to spend for one physician, while charging the patient a fee that is only minimally discounted below the fee for seeing a physician. Although the gap between salaries of these physician extenders and salaries of primary care physicians is closing rapidly, medical management business corporations and the schools in the business of producing physician extenders are still quick to point out that a difference does yet exist; thus they argue that utilizing physician extenders saves money. They are absolutely right. It does save a minimal amount of money— for the hospitals and physicians employing them! The patients, however, receive only a much less qualified medical evaluation, often over-done with unnecessary testing, and at a charge approaching that for which they could have seen a physician. Frequently they even end up being charged by the physician extender for the *service* of being referred on to an MD, when a primary care physician seeing the patient initially would have been qualified for diagnosing and competently treating the patient's problem on-the-spot.

I have known some extremely knowledgeable physician extenders, several of which I would be glad for my family to utilize if they needed medical attention and *if no competent physicians were available*. Some of the physician extenders I've worked with were formerly nurses in specialty units of the hospital, such as CCU or neonatology units. Some of them have even known more than I regarding specific techniques or technology utilized daily in that particular area of the hospital. None of them, however, possess anything near the overall depth of knowledge in all of medicine's disciplines, nor the diagnostic capabilities or total patient management skills that I do. This is not because I'm smarter than they are. It's because the education pathway I chose upon entering the field of medicine was the longest, most rigorous and in-depth of them all— the MD degree.

On one occasion I was asked by my hospital to provide clinical training and preceptorship for one of our staff nurses seeking to become a certified family practice advanced nurse practitioner. The individual had her masters degree in nursing, was highly motivated, and conscientious to a fault. At the completion of her preceptorship she would have been a dream working as an office nurse *under responsible and close supervision* of a family practice physician. As diligent and good as she was, however, it was evident the entire time of her training that she did not even think in the same clinical manner or at the advanced clinical level as I. This was not her fault. It was not possible for her to do so because she had not had the years of exhaustive basic medical sciences education that culminates in the scientific, physiological approach to diagnosis and treatment utilized by all good physicians. It's this mechanistic, physiological, mindset towards diagnosis and treatment that all traditional medical schools' curricula are geared toward instilling in their physicians. Medical schools strive for this, since this unique assessment approach is what enables a physician to recognize and successfully manage conditions he/she has never seen before— unexpected conditions, conditions presenting atypically and conditions occurring concurrently with other illnesses. It is a unique manner of looking at a clinical problem and the necessity behind the arduous educational requirements for the MD degree. It can't be memorized or copied, and there is no short cut to it! It is pure physiological, anatomical and biological sciences processed by a physician's mind into a logically scientific approach towards diagnosis and treatment. Simply put, only by knowing in extreme detail the processes necessary for physiological life and good health, can one understand where in that process things could go wrong, resulting in the abnormal condition characterized by a particular illness. Over and over during my serving as preceptor for this primary care nurse practitioner, I was amazed at the difference of the levels from which we looked at things. Her approach was superficial, based primarily on a memory approach and dependent upon her recalling clinical problems or cases she had seen an instructor or physician manage on a past occasion. She usually addressed satisfactorily the minor and routine problems that she had a past occasion for observing a physician treat, but always in discussion she would let something slip that told me she had not a clue as to why the treatment worked or how the illness's characteristics developed as they had. This told me that the unique, unusual, unexpected, atypical or cases involving multiple illnesses and multiple medications would likely get her into trouble eventually. Education via parroting and mimicking is not sufficient for the fluid and constantly changing playing field of clinical medicine. This is the reason that physician extenders should never be used in lieu of primary care physicians as first-contact providers i.e., for seeing a patient before the patient has been appropriately screened and an accurate diagnosis made by a physician.

For the vested interest parties intent on trying to defend the quality issues of physician extenders practicing with little supervision or independent of physicians entirely, I ask them to simply remember the golden rule and consider a hypothetical scenario in which they receive a phone call from their sixteen-year-old son vacationing in Florida. He informs them that he is having violent headaches, chills, vomiting and a temperature of 104. He also reports to them that the small town he is in has only one young physician and one young primary care advanced nurse practitioner, each with their own individual practices. Which of these two medical resources would they advise their son to see? All advanced nurse practitioners and other physician extenders should ask themselves this same hypothetical question before deciding to enter these fields. Ah, the *golden rule*, so effective and so conspicuously absent in most areas of medical care today.

In the extremely rare geographic areas today that remain without a physician, if indeed any still exist, the presence of well-trained physician extenders could be beneficial, but the scattering of them throughout the doctor-glutted urban areas and clinics is indefensible and the motivation for it strictly financial. I personally know of an ARNP who, prior to becoming a primary care nurse practitioner, was an outstanding and highly valued ICU/critical care nurse but now, as a primary care nurse practitioner, she is practicing medicine in her own "women's clinic", with emphasis on cosmetic procedures, including lucrative Botox injections! That anything other than self-serving special interests could be responsible for selling the nurse practitioner concept as beneficial and cost efficient, is difficult to believe in the extreme. Any person of common-sense, upon familiarizing themselves with the primary care ARNP programs' training requirements; their minimally or totally physician-unsupervised natures; and their open-ended scope of practice privileges must angrily ponder how such a perilous dupe of the healthcare consumer could possibly have been permitted to occur. It defies any reasonable logic that a program whose curriculum is designed by nurses, taught by nurses and completed in three to four years, with much of the credit obtained in community college class rooms or on line, can produce clinicians the equivalent of MDs with the latter's' four years of college, four years of intense medical school sciences and three years residency specialty medical training. Think about it!

The motivation, development and pitfalls inherent with the primary care nurse practitioner programs can be better understood after considering the following hypothetical commercial airline pilot analogy. Reading in an inaccurate news article that 100,000 commercial airline pilots will be needed by 2012, an enterprising flight controller recognizes an opportunity. Having political connections, a strong lobby available to him and an entirely naive state legislature regarding commercial airline pilot training; he is able to get legislation enacted for approving the use of commercial-airline-pilot *technicians* who, according to him, can be educated quickly given their experience in the airline industry as former stewardesses and air comptrollers. Additionally, he convinces the pilot-naïve legislature that commercial airline technicians produced by his program can function just as well as traditionally trained commercial pilots. An airline stewardess, dissatisfied with her job and interested in a better income, inquires how this new commercial airline technician program works. She is told that she will receive credit for her air-time as a stewardess, and the rest will be taught her by the new program's training faculty. "What does your program's faculty consist of?" she queries the new school's representative.

"Our teaching staff is made up of four former stewardesses, each having his or her recreational prop engine pilot's license," he explains to her.

"You're telling me that if I complete your program's requirements I will be legally able to fly commercial jet passenger planes for a living," she asks incredulously.

"Absolutely, see *State Statute 110-Air-Commercial Pilot Technicians,"* she is assured by the program representative. Still not convinced, the potential student persists,

"Well, where does the experience flying large commercial jet crafts come from? Your faculty only has recreational single-prop-engine experience," she persists. Now quickly warming to his sales-task, the program representative responds,

"The last two months of the program are spent riding in the cockpit of a commercial passenger jet liner where you get to observe the pilot at work and ask any questions you wish." Believing this too good to be true, the prospective student probes further,

"And that's all there is to it? Then I can be employed as a commercial jet airliner passenger pilot?" Closure at hand now, the program representative admits,

"Well, there is one more legal requirement. You have to get a licensed commercial airline pilot to sign a *collaborative agreement* with you. Suspicious all along, the stewardess lets the representative know,

"I knew there would be a catch to it! This is where I get assigned whatever aspects of the flight the pilot doesn't like doing or only the very simple tasks he thinks I can handle."

"Not at all," the representative assures her. "The collaborative agreement only stipulates that the educated and experienced commercial pilot entering into the agreement with you will always make himself available for consultation anytime *you decide* that you might need help or consultation. He doesn't even have to be in the same plane with you. You can pretty much do your own thing, and it's up to you to decide if you might need to consult with him."

"One more thing then, will I have to pass the same test that commercial pilots from traditional schools take?" Confident now that he has recruited a student, the representative's reply comes quickly.

"Oh no, that's another advantage of our fast track program. You only have to pass the commercial airline technician test, and our program designs its own test."

Even worse, it appears from my experiences that the primary care nurse practitioner programs in my area of the country are already evolving in the same negative fashion as the 2-year associate RN programs did.

Initially, the 2-year associate RN programs were extremely selective with their admission applicants. Only the brightest and most highly motivated were accepted. Also, in the beginning, 2-yr associate RN programs were located only in larger communities, frequently as an additional pathway in already established BS nursing program colleges. Clinical skills training and medical science education in the new 2-year programs, although short-changed in comparison to the traditional 3-year programs, Bachelor programs and Master RN programs, initially were compensated for by meticulously selective applicant screening and by having carefully selected teaching faculties that were highly motivated by their desire for the new programs' success. As graduates of these first of their kind 2-yr programs were quickly snapped up by our nation's rapidly expanding hospital industry (now under the guidance of big business corporations) at unreasonably high salaries for a two-year educational process, word quickly spread of this lucrative new professional title, and applicants flowed out of the woodwork for taking advantage. Never mind that their four-year-degreed colleagues and medical staffs within hospitals recognized immediately how short-changed these new comers' clinical skills and medical science backgrounds were, compared with their three and four year degreed predecessors, students flocked to the 2-yr program schools. This fueled a demand and with that demand came business opportunities for more two-year programs, and, in no time, every tiny burg with a technical school or junior college jumped on the band wagon. Never mind that many of these schools lacked faculties for adequate medical science education or medical facilities large enough for providing adequate clinical experiences. Churning out numbers and keeping the new schools' funding intact became the new goal. For pursuit of this goal a large and continuous supply of students was needed, and the means to this end appears, in many instances, to have been a relaxing of the late-comer programs' admission standards, unbelievably so in many cases.

Several years old now, many of the primary care ARNP training programs appear to me to be following the same declining quality in their evolution that the two-year nursing programs did two decades before them. It's now become all about numbers and the survival of the programs themselves. Increasingly, primary care ARNPs are appearing to me to come out of these programs possessing significantly less medical science backgrounds and fewer clinical skills than the three-year-degreed nurses staffing our hospitals thirty years ago possessed. Indeed, it is becoming frequently more noticeable, through observation and interaction with them in clinical matters within the hospital, that it is often difficult to differentiate the more recently trained primary care nurse practitioners today from their two-year associate degreed nursing colleagues. Despite this fact, today's primary care ARNPs have legally been given significantly more clinical privileges and independence, essentially enabling them to practice medicine and, in the process, adding additional and extremely economically inefficient hands into the nearly empty reimbursement cookie jars of Medicaid and Medicare.

In the sixties and seventies we had our own version of nurses with expanded clinical roles and tried to staff the eleven to seven ER shift with them in an attempt to reduce our trips out to the hospital during the night. Back then, though, they were called "nurses", the three or four year degreed types who *were* sufficiently grounded in the medical sciences for communicating and assisting physicians and for serving as our proxies with some phone supervision. It was testimony to how much better these old-school nurses' medical science backgrounds were that nurses then, possessing enough medical science education for recognizing how superficial their education and medical science backgrounds were compared to that of an MD, had to be begged and cajoled into serving in more expanded clinical roles.

Once persuaded to accept more expanded clinical responsibilities, physicians in those days carefully delineated, for these carefully vetted nurses, a limited number of expanded clinical tasks that they knew them to be capable of performing well. Even then the physicians monitored the performances of these select nurses carefully and continuously, since they understood all too well that they were ultimately responsible for these nurses' actions. The collaborative agreement being entered into between physician and primary care nurse practitioner today, however, has no meaningful supervisory safeguards for patients and no mechanism for making the physician responsible or, for that matter, even feeling a shared responsibility for the actions of the nurse practitioner; thus, there is minimal to no meaningful on-going physician supervision occurring in most instances today.

The area in which a consumer and medical paradigm shift could impact healthcare delivery the greatest, from an economic and quality care perspective, is in the delivery of end-of-life care. Medicare studies have shown for decades that the bulk of Medicare dollars go to the last few days or weeks of patients' lives, usually in a futile and misguided attempt at prolonging the inevitable. No other area in medicine could benefit more from application of the *golden rule* than end-of-life care throughout our nation's nursing homes and skilled care facilities. Providing the general public and nursing home staffs education regarding the difference between old-age and old-age in end-of-life circumstances due to multiple end-stage-disease processes is crucial if we are going to economically survive the aging baby boomers over the next two decades. In particular, the public desperately needs a clearer understanding of the difference between quality life and physiological life. The decision to *pull the plug is* seldom a relevant quality care issue for patient or family today, but the doctor's ability for *plugging in* today's expensive technology for maintaining meaningless, physiological life is a daily quality of care issue throughout our nation's hospitals and nursing homes.

Sadly, our physiological, life-sustaining, technical advances have followed a similar evolution in their misapplication as CPR (cardiac pulmonary resuscitation) did. Misperception by the public, dramatization by the media and fear of the legal profession and regulatory agencies has resulted in the medical profession's widespread misuse of a concept and modality originally developed for and effective in only a very few select clinical situations. CPR, developed primarily for intervening in acute cardio respiratory failures resulting from drowning, electrocutions, and acute cardiac events in otherwise healthy individuals, has been generalized to the point of becoming the norm whenever clinical death occurs, regardless of the circumstances. That there is an expected and proper time for death is being ignored entirely in our nation's nursing homes and hospitals. CPR usually is applied to some degree regardless of whether it is wanted, indicated or justified. Given the fact that the process almost never leads to any meaningful prolongation of life when utilized for patients having multiple end-stage-disease processes, scientifically and statistically, the procedure actually should be applied as the exception rather than the rule. Education of the public could begin with changing our way for ordering no application of CPR. Presently we use the order *do not resuscitate* (DNR), implying that we have the ability to always resuscitate all cases. The daily entertaining but inaccurate portrayal of CPR by popular television medical dramas is misguiding the public in the extreme. In actual fact, positive outcome statistics are sobering indeed, and the order for withholding the procedure would be much more accurate if written "do not *attempt* resuscitation." Physiologic life-sustaining technology such as respirators, dialysis machines, feeding tubes and vasopressor drugs have become the rule as the organ systems normally responsible for maintaining these functions begin to fail, without any consideration whatsoever to the overall context in which the failures are occurring. This technology was developed for utilization in specific circumstances where their need was expected to be short term and where the patient could be expected to return to a reasonable quality of life and longevity, not for patients having multiple organ system end-stage-diseases in their last days to weeks of life. The respirator that is a miracle for a young overdose or asthma victim becomes a hellish nightmare for an end-stage emphysema patient, in the hospital more often than out for the past five years of his life or for the nursing home patient with multiple organ system end-stage disease and no meaningful quality of life nor expectation of any in the future, regardless of medical treatment rendered.

Like the elderly Ms. Jarvis whom I met at "the old man's" bedside many years ago, we all well-know who these patients are and that application of the *golden rule* is best and sufficient for guiding us toward decisions in their best interests. Yet, it rarely is done timely or properly. For physicians, the primary excuse would be fear of inappropriate and undeserved litigation. They argue that, when terminal events begin, it's better to error on the *try*-side than to offer comfort and dignity over futile, physiological, life-support at the end. Physicians always fear that some family member, not present at the time, will turn up later and sue for not doing everything possible for avoiding death. In opting for comfort and dignity measures for their end-of-life patients, physicians also worry that some less-informed bystander, such as nursing staff or ancillary medical services staff, might misinterpret such actions and, in so doing, cause needless trouble for the physician by perpetuating their misperceptions through in-house gossip or through the paper trail. Even more likely to be the concern of physicians today is, by opting for comfort and dignity measures, they might inadvertently violate some portion of the paper trail mindset and, as a result, find themselves before some in-hospital committee or regulatory agency to defend their actions.

The best remedy for the first of these scenarios is advanced communication, if not actually by an advanced directive, at least by knowing the patient's wishes and patient's family's wishes from good communication over the years. Unfortunately, this kind of doctor/patient and doctor/family relationship, so valuable to all parties when medicine was still a profession and each patient being managed by a primary care physician, is all but extinct in today's medicine as a business. A great part of the problem today is that, once in an acute care hospital, a patient usually will be seen by several physicians instead of one, and communication among them all is usually bad. The recent trend of *hospitalists* (physicians only providing care while the patient is in the hospital) replacing patients' personal primary care physicians as coordinators of the general medical care in hospitalized patients has simply added to the problem, since hospitalists have no more personal knowledge of the patients' medical histories or their end-of-life preferences than any of the other physician-consultants briefly seeing the patient in the hospital setting. For example, a pulmonologist, requested by a hospitalist-physician, only in the interest of the patient's comfort, to consult on chronically low oxygen levels in the blood of an 85-year-old end-stage-dementia patient with pneumonia, might be at the bedside when the end comes and place the patient on a respirator. If asked why, the usual answer is, "Well, it's not my patient, and, not knowing their history, I felt it best to go this way." Best for whom, certainly not the patient undergoing the discomfort of the procedure and now being relegated to an additional several days suffering before the inevitable occurs and certainly not to the family who will now get to learn first-hand that it is far easier to let a loved one go naturally, sans technology, than it is to be specifically asked their permission for "turning off the machines" after the fact.

For medicine as a business to be more economically friendly and patient focused, the Joint Commission's paper trail paranoia within hospitals and nursing homes must be exposed for what it is and replaced by a meaningful and cost efficient regulatory inspection process that emphasizes patient outcomes, patient and staff interviews and unannounced on-site observations. The cost in overhead of the current paper trail mindset within hospitals and nursing homes today is so astronomically high that even if it guaranteed quality care, which it certainly has not, and the public ever learned how much they were paying in increased charges due to it, they would clamor for an alternative. We need to do away with the policy-review orientated, hospital-resources gobbling and budget-busting Joint Commission.

Once Joint Commission is taken care of, attention should be directed to the states' Offices of Inspectors Generals (OIGs) and serious consideration given to modifying entirely their approach to regulation and inspection as well. The goal of changing states' OIGs' ways of doing business would be to produce a more properly motivated, outcome orientated, clinically-grounded, real-world inspection team. Proper motivation would have the inspectors eager to impact positively on outcomes in services, rather than focusing their efforts toward compilations of so-called violations which, in and of themselves, have little to do with patient outcomes or the day-in-day-out patient care quality within a facility. For effecting such a change, people on the team would necessarily have to be clinically contemporary, extraordinarily motivated, people friendly, realistic and innovative thinkers. The only source for such individuals is the best and most successful of those currently working in the healthcare field. For maintenance of their motivation and positive attitudes, it would be necessary that their terms of service be limited, perhaps to one or two years. Limited terms would prevent burnout, assure contemporary skills and assure all their efforts being directed toward their jobs and not toward job security, as is so often the case today. Most importantly, limited terms in OIG would assure that the individuals reviewing the facilities, having recently "been there and done that", would be realistic in their solutions. For luring these successful folks from their current jobs, a very lucrative salary for their stint of service could be offered. It should be attractive enough that healthcare facilities could use this as a built-in competitive performance challenge. Publicity ground work could be laid throughout the medical industry so that it became common knowledge that, if you were thought of as one of the brightest and best in your field, you would have a chance of being offered a two-year leave of absence to do your state and profession a service, while being remunerated handsomely for your time. Most important of all, there should be MD representation on every OIG inspection team! It is indefensible and defies common-sense that physician representation has been omitted from an inspection entity whose sole purpose is judging the quality of health care delivery!!

Also, if healthcare costs are to be lowered or maintained at current levels, one of the major cost-spiraling forces, defensive medicine, must be eliminated. For physicians to give up the defensive medicine game and resume ordering ancillary studies by the rules and concepts they were taught in medical school, they must be freed from the constant fear of unjustified lawsuits. One possible way for this to be addressed would be pretrial committees composed equally of physicians, laypersons and judges for reviewing the merits of any malpractice suit and eliminating those without merit from being tried. Caps on pain and suffering would also help enormously. To be effective, any torte reform must be swift, dramatic and leave no doubt in physicians' minds.

Healthcare costs and the efficiency and quality of today's medical business would benefit extremely from a properly administered gate keeper concept in one form or another. Properly administered, in this instance, means the gate keeper position could only be held by a physician having broad but in-depth training and experience in all the medical disciplines. General internists, family practice physicians, general pediatricians and emergency medicine physicians all fit that bill nicely. For reassuring those patients who might feel they are being kept from a procedural specialist unwisely, rapid and free second and third opinions could be arranged by referring the patient to two other gate keeper physicians in unrelated organizations.

Frontline triaging of patients, as required in the gate keeper role, is the most important, difficult and high-risk task in all of clinical practice and requires a physician to draw on his/her extensive education and training more often than any other area of medical practice, for differentiating the routine from routine-mimicking life-threatening catastrophes! Under no circumstances should this vital responsibility be left to a physician extender as is so often being done today in busy clinics and emergency rooms throughout the nation. If, however, physician extenders continue to be utilized for a while longer in the absence of sufficient primary care physician numbers, it is much more sensible and safer having a primary care MD initially see the patient for triaging their complaint and degree of illness. Should the patient's complaint be the "worst headache of my life", then, after the MD has ruled out brain tumor, encephalitis, meningitis and brain hemorrhage as its cause, the patient could then be referred to the physician extender for counseling and treatment of her migraine or muscle contraction headache but, even then, under consistent and meaningful supervision of a physician.

An area crying out for improvement today is nursing education. Over the past thirty years it has ceased being based on patient, nurse or physician needs; it now dances entirely to the music provided by the duet of business demands and financial gain. Their profession, like mine, is dead. Its cause of death was an increased demand for nurse numbers sufficient for medical management corporations' planned expansion of services in their efforts to increase their bottom lines. Medical management corporations, common by the mid-seventies, solved their need for increased nurse numbers by supporting the concept of the educationally diluted two-year nursing degree programs. Experiences with recent graduates of the two-year programs in my area of the country lead me to suspect that these programs have cut their nursing basic medical science education to the bone, and their hands-on clinical experience is non existent in far too many instances. My personal impression is that the two-year RN associate programs emphasize nursing philosophy, record keeping and legalities, while giving only a superficial glance at medical sciences and clinical techniques. The focus in these last two areas seems increasingly to be on that which will be asked on the boards for licensure testing. For the price of tuition and only two years of their time, they receive an educationally inflated title and the legal right to employment by hospitals at salaries far out of proportion for their education, abilities, experience and worth to the system. Nurses themselves will admit to you, as they have to me, that some of them turned out of such two-year programs in recent times have never given an IV injection to a patient, assisted in a sterile office surgical procedure or inserted a Foley catheter. Many doctor colleagues have shared with me that a great many of the two-year nurses being churned out today lack medical science background sufficient for even effectively communicating with doctors. It would be safer for patients and better for physicians and the nursing profession if the two-year school graduates were confined to paperwork issues, calling them *clerical RNs* and scaling back their pay accordingly. Today, it is likely that nearly seventy percent of all nursing new hires in our nation's hospitals, clinics and nursing homes are from the 2-year associate RN degree programs. Actual clinical assessment and clinical decision making for the patients should be returned to those now working as advanced registered nurse practitioners (minus their diagnostic and prescriptive privileges) and other higher-degreed nurses. Doing so would place hospital and nursing home patients back into nursing hands qualified for assessing and monitoring their clinical conditions. Redefining the rolls and job descriptions of ARNPs would bring them back to a job within the system that is sorely in need of increased numbers and improved quality. Even better, it would be returning them to a job for which they would be appropriately qualified for performing well.

Additionally, it would go a long way toward solving the quality, experienced-nurse shortage within our nation's hospitals, since another predictable and undesirable result of permitting BS degreed nurses to have prescriptive authority and assume surrogate physician roles, along with near physician compensation, has been a *brain drain* from the nursing departments of our nation's hospitals.

For physicians claiming that their practices have become income-dependent upon an ARNP's services, they could be permitted continued use of them but in a relationship more compatible with quality and safety. The agreement signed between ARNP and the physician should specifically assign the physician equal liability for all acts and duties performed by the ARNP. Additionally, instead of open-ended prescriptive authority, the ARNP should only be permitted to write for a limited number of specific drugs as approved and agreed to in advance by the physician entering into the agreement with them. Also, the ARNP could only legally prescribe for the drugs on that list during the course of his/her work while physically present in the physician's offices or clinic and then, only if the physician were present in the same clinic at the time it was written. This would provide the consumer with the benefit of much-needed real and meaningful supervision by a physician.

Should we persist in accepting that the two-year associate degree nursing programs are capable of producing ready-for-the-work-place nurses, with clinical education and skills comparable to the three-year nursing programs and BS degree programs of the past, at the very minimum, effective oversight of the numerous and prolific facilities producing them must be undertaken. This oversight should emphasize quality and standardization among the many schools, particularly in regard to faculty qualifications, student-applicant quality and availability of appropriate clinical experience venues i.e., medical facilities and hospitals of sufficient quality and size for providing clinical training and clinical experiences.

For medicine the business to be economically more consumer friendly, a definite need must be met for the public's education regarding the difference between their realistic medical needs and their perceived medical needs. Additionally, they need to be made aware of the role that they themselves inadvertently play in the high cost of health care delivery today. Towards this end consumers need to become aware of their own contributions via frivolous, opportunistic litigation, and they need to become cognizant of the media's favoritism toward technology and *boutique medical practices* over the value of sound clinical judgment from a well-trained and properly motivated primary care MD. They also need to understand that self-referral to non primary-care specialists, especially procedural specialists, is seldom a good idea. Even if their self-diagnosed problem turns out to fall within the scope of practice of the specialty they choose, they frequently wind up paying a much higher fee for the same treatment that a primary care physician could have provided them. Obviously, if a patient self-refers to a general surgeon for abdominal pain and that pain should turn out to actually be due to appendicitis, things work out very well for both patient and surgeon, but, even then, a primary care physician in the same medical community might have made the referral of the patient to any of three other general surgeons in the community as a first choice, knowing that the surgeon the patient chose out of the yellow pages was nearing retirement and no longer doing enough surgeries per month for keeping his skills up or having any of a number of other personal problems or professional ones that would have prevented him being the primary care physician's first choice for a surgeon. Directing the consumer to the appropriate specialty and selecting the best available person within that specialty are a primary care physician's responsibility and another extremely important reason for all patients having a primary care physician. Primary care physicians oversee their patients' total health care needs and assist in guiding them through an increasingly complex and patient-unfriendly medical system today.

By far the greatest reason that self-diagnosis and self-referral often wind up being detrimental to the patient and to the cost of care issue as well, is that there are far more medical conditions causing abdominal pain that do not require surgery as treatment, than there are those that do. In this hypothetical example of self-referral to the general surgeon, it is overwhelmingly likely that the surgeon, after briefly interviewing and examining the patient, would have said to him,

"I don't find anything that leads me to believe that you have a condition at this time that is a surgical one. Pain such as you describe could be caused by any of a number of medical conditions. Do you have a primary care physician that can sort through this for you?"

In the distant past, the majority of non–primary care specialty physicians accepted only referrals made to them by another physician. This gradually began to change around the early eighties and begs the question, "why?" After all, if non–primary care specialists were sufficiently busy and financially happy plying only the specialty techniques and skills of their self-chosen limited fields, why would they feel the need to see greater numbers of patients, especially if those patients might turn out not to require their specialty techniques or skills? Greed is always a potential possibility but the switch by procedural specialties to accepting self-referred patients has become too widespread for greed to be the only factor. A much simpler and plausible reason might be that, today, focal specialists increasingly are finding it difficult to stay busy enough if they confine themselves strictly to seeing only patients who unequivocally have need of the expertise of their particular specialty. Ironically, medicine over the past thirty years has been co-opted by business in every principal but one, supply and demand. Currently it would appear that supply of many of the high-profile, high-pay specialties is now exceeding bona fide needs for their services. If so, this creates the ideal conditions for producing an increase in unnecessary procedures and surgeries and, for the same reasons, creates a temptation for sub specialist physicians, trained in very focused areas of medical practice, to begin dabbling in areas of medicine unfamiliar to them in order to remain sufficiently busy.

Those old enough for doing so would remember the fifties to the early eighties being a period characterized by high patient satisfaction, high doctor satisfaction and economic stability within our nation's health care system. It was also the time frame in medical history where primary care physicians were the work-horses of our health care system. During this period they far out numbered their specialty colleagues and provided competent, efficient and cost effective care for the lion's share of all of America's injuries and illnesses. In doing so they also identified and directed the small percentage of patients, in need of a specialty other than primary care, to their most respected and reliable colleagues within the appropriate specialty for dealing with the particular problem. In short, primary care physicians, serving as gate keepers to the nation's health care system of the past, provided a much needed service for their patients, their specialty colleagues and *the economics of our health care system.*

The abandonment of the effective and efficient medical referral system of the past began with the advent of Medicare, in 1965, and its reimbursement formulas for administering fees for services to physicians participating in the program. Reimbursement was higher for specialty services, particularly and disproportionately so for specialty services involving surgery or other invasive intervention or instrumentation. Private medical insurance companies were quick to adapt Medicare's reimbursement strategy, and, so, as long ago as 1965, the seed was sown in two separate gardens, that of the general public's minds and within the minds of the medical profession. These planted seeds would ultimately germinate into the public's concluding that if it costs more it must be better and into our nation's medical students concluding that specialties other than primary care lead to higher incomes.

However, Medicare's singling out specialists and invasive procedures for disproportionately higher reimbursement fees was not the only factor propelling the health care system into the specialty orientated one of to day, where health care consumers subscribe to the anatomical-specialty mindset, deciding for themselves the type of physician they require for their perceived problem and then using word-of-mouth or the yellow pages for finding them. Of course there were other forces influencing the trend as well. In fact, the sixties and seventies provided a "perfect storm" atmosphere for assuring this would occur.

Birth of the large business corporations specializing in the management of hospitals, an explosion in the development of new technology and the media's sudden fascination with everything-and-anything medical played very significant roles as well. Prior to hospital management by business corporations, the nation's hospitals (with the exception of those operated by large charity organizations) were managed by a hospital administrator who answered directly to a local hospital board comprised of successful and public-service-minded local residents from various walks of life within the community. The administrator was responsible for all aspects of the daily operations of the facility, while the board was primarily involved with long-term management of finances, long-term goals and maintenance of a suitable physical plant. Completely unlike the circumstances within hospitals' today, in all matters regarding patient services and quality of care issues in those days, both the administrator and the hospital board looked to the hospital's medical staff for guidance. In those days a noble and proud medical profession still existed, and medical staffs of those times consisted of tightly knit professional groups whose only interests were excellent patient outcomes, patient satisfaction and maintenance of their own professional reputations among their peers. Most of the direction from the hospital board was altruistic in nature and designed with the community's best medical interest in mind. The medical staff in those times, having a much greater voice and influence within the hospital venue and its daily operations, served much more vigorously and effectively as quality and safety advocates for their hospitals and for their patients. It requires little imagination for predicting potential quality and patient-dissatisfaction issues that would occur in switching from hospital administrations whose primary goals were quality of care, community service and professional reputation, to hospitals run entirely on the business model and whose primary concerns are healthy bottom lines for investors.

It was undoubtedly improvement in the bottom lines of their smaller rural hospitals that medical management corporations had in mind when they began recruiting types and numbers of specialties that heretofore had never been available to smaller rural health care markets. In their rush to expansion into the smaller markets and for making their smaller hospitals more competitive in specialty services being provided by nearby urban medical centers, hospital management corporations vigorously recruited specialists of all types, frequently using lucrative pay packages in doing so. This created a spurious demand for more specialists, followed quickly by complaints of a specialist-shortage whenever hospital management corporations couldn't find the numbers needed for meeting their expansion goals. The management corporations' specialist-shortage complaints, picked up on by the media and reaching the ears of the nation's medical students, provided the latter very interested parties with additional reason and motivation for choosing a non–primary care specialty upon graduation. Ignored entirely in this smaller market specialty services concept, however, was any consideration of statistical prevalence data per given population for illnesses that would genuinely require a particular specialty service.

Simply put, if a urologist is being recruited for a hospital in a community whose population is forty-thousand, and two other well-established urology specialists are located within a half-hour's drive from that hospital, how many illnesses and procedures requiring the expertise of a urologist will a population of forty-thousand statistically be capable of generating? Data exists for accurately estimating this answer, and certainly a recruited urologist would benefit the most from knowing the answer, but potential patients also have need to know that their urologist will be able to remain busy enough for assuring he is performing the requisite number of procedures annually that his specialty organization has deemed necessary in order for him to maintain his surgical skills. Similarly, the health care system's economy has need to know that the urologist is not going to have to resort to seeing routine urinary tract infection cases and charging urological specialty fee rates for doing so in order to earn the average income expected by urologists nationally.

The idea that specialty medical practices require larger population bases for financial viability than primary care practices is nothing new. In the sixties, the American Medical Association provided young physicians entering practice, free of charge, a lengthy pamphlet full of helpful information for initiating a private practice. This pamphlet, in addressing potential locations for a new practice, recommended the population deemed necessary for supporting a practice, listed according to specific specialty. The precise figures have long since been forgotten, but the practice type needing the lowest population figure was a general practice in primary care, requiring somewhere around 2,500 people per individual physician. Specialty practices of any other type required considerably larger populations, and the necessary population requirement for each climbed more steeply as the degree of specialization became more specific. This is not that difficult to understand. A doctor with qualifications for treating eighty percent of all medical conditions occurring within a community of 5,000 will remain quite busy, a doctor with qualifications for treating only fractured bones would be much less busy and a doctor limiting his/her practice to vascular surgery could not survive with only a population of 5,000 to draw from. Interestingly, the last and only other time the author remembers seeing this pamphlet containing the recommended population-per-specialty figures was sometime in the seventies, some ten years after first seeing one in the sixties. In only ten years time, however, the population figures deemed necessary for financial viability had mysteriously become significantly *lower* for all categories!? It is not known whether or not the AMA still tracks recommended population data today, but it would be very interesting to see current figures, for if they continue to document a downward trend, it begs the question how this could be possible. There could be reasons not readily evident at first glance, but common-sense would certainly lead one to suspect that the general public's anatomical-specialty mindset, increased numbers of proceduralist specialists and over-utilization of services might be playing a significant role. Anecdotal evidence supporting this includes: the majority of U.S. medical students today eschewing primary care in favor of the higher paying specialties, the continuing trend of specialties being available in smaller and smaller medical markets and the increase in TV, newspaper and bill board advertising by the high-pay specialties. Today it is not uncommon to encounter an interventional specialty practice in a community so small that, twenty-five years ago, it supported only two primary care physicians!! What does this tell us about non–primary care specialty numbers, and, more significantly, what does it tell us about the possibility of *over-utilization*? It appears likely that we are dangerously close to an over-supply in many of the high-pay and high-profile specialties, certainly we are if these sub specialty physicians intend on seeing only those cases clearly in need of their sub specialty expertise and not seeing cases of primary care nature that they no longer keep abreast of. Keeping the future supply of proceduralist-specialty physicians commensurate to bona fide, valid proceduralist-specialty statistical needs is an important quality of care issue and a necessity for the economic viability of the nation's health care delivery system.

Most surprising, for those outside the medical field, is that non–primary care specialists, engaging in medical diagnoses or treatments not routinely a part of their limited-focus practices, often do not address such problems as effectively or as efficiently as would their physician colleagues in the primary care specialties. This is particularly so for patients having multiple medical conditions and taking multiple medications. This is not that difficult to understand. A dermatologist, general surgeon or orthopedist is not likely to perform nearly as well in selecting a third antihypertensive drug for adding to the regimen of an elderly, hypertensive diabetic with stable congestive heart failure as would a physician in primary care. This is why it is imperative that we restore primary care *MDs* to sufficient numbers and sufficient remuneration within our health care system.

Not as surprising for those outside the medical field but certainly of extreme importance, is the economics of seeking out specialty care for addressing a problem that could have been addressed just as well by a primary care physician. Basic first time office visit charges of non–primary care specialists are almost always very significantly higher, even if no specialty techniques are utilized during the consultation. How much economic sense would it make for a young healthy female with honeymoon cystitis (a common bladder infection occurring in newly wed women) choosing to see an infectious disease specialist for the problem? Her problem is a bladder infection and, as such, certainly technically fits into the scope of an infectious disease specialty practice and likely would be addressed effectively and efficiently in this venue, but the fee charged by the infectious disease specialist would likely be two to three times the amount her primary care physician would have charged for the same effective treatment.

The point is that currently many of the high-pay, procedure-orientated specialties are becoming over-crowded when measured against bona fide valid statistical need for their services, where need is defined as the number of cases definitely requiring their particular area of expertise or technique for achieving a significantly better outcome than could have been achieved by primary care alone. Yet the evidence suggests today that the ratio of primary care specialty physicians to non–primary care specialty physicians is heavily weighted toward non–primary care specialists. This is an irrational and expensive direction given the quality and economic issues the system is currently struggling with.

It is difficult to understand the media's and consumers' constant over-valuation of the specialties involving surgery, invasive procedures or use of high-tech equipment and their constant under-valuation of the thinking, planning, guiding, preventative and holistic specialties in the overall medical scheme of things. The critically important role of primary care physicians to the efficiency, effectiveness, quality and cost control of our nation's health care system cannot be overstated!!!

Suppose you were planning on purchasing and living in a two-million-dollar home which had been standing empty for a decade. It is unlikely you would choose a carpenter or plumber for carrying out an overall inspection of the entire structure for determining needed repairs and upgrades. Good judgment would dictate your hiring someone with an in-depth knowledge of all components of the dwelling, likely an experienced general contractor or a home inspector. Either of these individuals would be capable of diagnosing problems within any of the home's components and making efficient and effective recommendations for putting right any problems found. Alternatively, you might feel you would be better off obtaining the appropriate specialist for inspecting each individual system of the home and hire a plumber for the plumbing, an electrician for the wiring, a roofer for the roof and so on. The cost of carrying out the inspection in this fashion would be exorbitant, and, in searching out the individual specialists on your own, you are flying blind regarding the competency or motivations of any of them. Primary care physicians are the general residential contractors of the health care system and much more. They are also the experienced, multi-talented general handymen and women of the health care system, competent for effectively repairing a very high percentage of your health care problems and possessing the overall knowledge of the health care system necessary for finding for you the best possible persons for addressing any of your health needs definitely turning out to require specific and expert services beyond primary care.

Again, the fifties through the early eighties was a time period when our nation's health care system was characterized by patient satisfaction, physician satisfaction, high quality and reasonable costs. It is more than coincidence that, during this same period of time, primary care physicians were the majority work-horses and the glue which held an efficient and effective health care system intact, and yes, they served as *gatekeepers* to the specialties as well. It is also more than coincidence that, during this same time period, the majority of physicians in practice had been trained and continued to practice according to the standards and ethics of a still-intact *medical profession. Sadly*, we cannot expect to ever resurrect the *medical profession,* but we cannot afford to fail in restoring primary care physicians to sufficient numbers, sufficient remuneration, sufficient pride and, most of all, to their effective roles as first-contact physicians necessary for accessing the many confusing choices and pathways available within our health care system today.

Accomplishing this will require a concerted effort from medical academia, primary care physicians themselves and the pertinent primary care specialty societies. A major responsibility of the primary care medical societies will be assuring continued maintenance of their training programs' current comprehensive academic exploration of all the major medical disciplines. Primary care specialty societies also will need to assume the difficult but sorely needed role of surrogates for the now extinct *medical profession,* setting the bar high for membership and then monitoring and disciplining that membership for: lapses in ethics, non adherence to quality standards, clinical decision making contrary to medical school scientific tenets or repetitive case management contrary to evidence based medicine. In short, each primary care society membership will need close monitoring from their society for recurring evidence suggesting that case management decisions are being made on the basis of self-serving interests such as convenience, defensive medicine or financial gain and for meaningful effective action taken against all repeat offenders. The percentage of physicians today engaging in case management decisions based on self-serving reasons, rather than the scientific medical ones they learned in medical school, is likely far greater than the most cynical estimator would suspect and, besides representing poor quality of care, is adding multimillions, perhaps, billions of dollars to the cost of health care annually.

If there remains any hope of our health care system becoming economical, efficient and effective once again, properly motivated and traditionally educated physicians of the primary care specialties must again become the first and last stop for the lion's share of patients accessing the health care system. Primary care physicians' training is specifically designed for their competently and safely serving in this capacity. Additionally, their education and training is especially well-suited for weeding out the much smaller number of patients in definite need of sub specialty care and for guiding these patients to the most appropriate level of specialty care needed and into the best possible of hands available at that level. The public needs enlightened to the fact that the education and training of physicians in the primary care specialties is just as lengthy and comprehensive as for other specialties. The major way in which their education and training differs from that of most of the other specialties, however, is that primary care training focuses on competency in diagnosis and treatment of illness and injury developing in any of the body's organ systems and, in particular, diagnosis and treatment of illness occurring in patients already having other pre-existing multi organ system diseases under treatment and already taking multiple drugs. Such patients are rife with potential for medical errors and complications, and only the training and education of the primary care specialties is broad enough and in-depth enough for managing these types of difficult patient profiles. Aside from the difficulty in interpreting the significance of signs and symptoms in patients where coexisting illnesses might be masking or exaggerating signs and symptoms of a new illness, adding an additional drug to the long list of drugs that such patients are already taking for other conditions, regardless how common the additional drug's usage or chemical class, can become an unexpected nightmare in this patient population. Prescribing medications for such patients has pitfalls too numerous for listing, and, even for seasoned primary care MDs, this patient type is extremely difficult to manage.

Pharmacology considerations have become so complex today that most conscientious and honest MDs would likely admit, as they have to me, that the semesters of formal pharmacology classes labored through during the medical school years are no longer sufficient, and, as is the case for all the basic sciences of the medical school years, honest and conscientious physicians wish they could go back for more pharmacology and physiology. For such highly-educated physicians to feel a need of additional pharmacology knowledge beyond that received in their medical school training is extremely note worthy, given that today we continue to push the envelope of prescribing privileges. Today, several allied medical health care providers already have been granted prescriptive authority with little likelihood that their formal training in biochemistry, physiology and pharmacology approaches even the tiniest fraction of that required in a traditional medical school for being granted the MD degree. The educational requirements necessary for general prescriptive authority is not just a topic for academic debate, nor is it based on turf wars. It is a very real matter of public safety and a real quality of care issue. In the practice of medicine, *simple cases* exist only in clinically *simple minds*. Conscientious MDs are trained to see traps and pitfalls in every differential diagnosis and with every medication prescribed— an educational point that I recently had the occasion for discussing with a new primary care nurse practitioner employed by the hospital.

Noticing how frequently and inappropriately this individual was prescribing the popular antihistamine/decongestant, Claritin-D, I broached the subject with her. Somewhat defensively she began her response by informing me that she limited her management cases to only *simple things* like "colds", "allergies" and "high blood pressure." Her response afforded me the opportunity for presenting the following hypothetical teaching scenario for her consideration.

A seventy-eight-year-old male comes to the drugstore mini clinic she is staffing and complains of four days of the worst stuffy nose he has ever had in his life. The patient further emphasizes the severity of his problem by adding that not being able to breathe through his nose at night has caused him to remain entirely awake for the past three nights. He goes on to explain that he suffers chronic insomnia, and now, not being able to breathe through his nose, has aggravated the situation tremendously. The nurse practitioner then asks Mr. Stuffy Nose for his past medical history and learns that the patient has a history of congestive heart failure, which, so long as he takes all four of the medications he was prescribed for it and avoids salty foods, has become stable enough that he can again walk on flat terrain if he doesn't hurry. Mr. Stuffy Nose also informs her that he has high blood pressure which has been doing "good"[sic] after the recent addition of a third antihypertensive medication to his regimen. The nurse practitioner then prescribes a ten-day supply of Claritin-D for this patient's nasal symptoms, pleased for the opportunity of being able to solve another *simple problem* with a popular drug that she knows to be in widespread use and which enjoys a relatively safe side effect profile. Yet, for this particular patient, any or all the following problems could result from his taking this popular medication. Claritin-D is a compound drug containing pseudoephedrine as the decongestant and loratidine as the antihistamine component. Pseudoephedrine belongs to a class of drugs called sympathomimetics, and drugs of this class have the ability to mimic the effect of sympathetic nervous system stimulation of organs under the influence of the sympathetic nervous system. They can accomplish this directly by activation of the sympathetic nervous system's alpha and beta receptors on the cell membranes within the target tissue or indirectly, through stimulation of the release of norepinephrine from the storage vesicles of the sympathetic presynaptic neurons supplying the tissue. By the latter mechanism, norepinephrine, when released into neuronal synapses, is then free to activate the post synaptic alpha and beta receptors. By either mechanism, it is the stimulation of alpha or beta receptors on the target tissue's cellular membranes that results in a specific response by the tissue. Armed with an intricate working knowledge of which tissues are rich in alpha and beta receptors and the expected receptor-specific response associated with a given tissue, it is possible to predict the physiological result which might occur in various organs as a result of being administered a drug like pseudoephedrine. As they say, forewarned is forearmed, and, had the nurse practitioner possessed a medical school education in sympathomimetic pharmacology and sympathetic nervous system physiology, she might have immediately sensed a red flag with regard to prescribing a sympathomimetic drug to a patient with this kind of medical history and taken pause for coming up with a better plan.

Already unable to sleep because of his pre-existent chronic insomnia being aggravated by nasal congestion, stimulation of the central nervous system's alpha receptors by the pseudoephedrine component of the Claritin-D stands a reasonable chance of assuring that Mr. Stuffy Nose will remain wide-eyed the entire night, since its stimulation of the alpha receptors within certain areas of the brain leads to a hyper-alert sensation that, while variable on an individual basis, is quite marked in a significant portion of the population. In plain speak, it wires them up.

Due to his advanced age, Mr. Stuffy Nose, while taking Claritin-D prescribed him for his nasal congestion, has a very good chance of experiencing an even more unpleasant problem, due again to the pseudoephedrine component of the Claritin-D. This particular problem will certainly be likely to make him forget about his nasal congestion for the moment. He may well begin having difficulty emptying his bladder. Indeed, he may even find it necessary to make a trip to his primary care physician's office or to the ER for an in-and-out catheterization for emptying his painfully distended bladder. Beyond the age of sixty, a very high-percentage of the male population already have an enlarged prostate gland and are experiencing some degree of symptoms with their voiding, or soon will be. Smooth muscle fibers within the bladder neck contain what? That's right, alpha receptors again! Stimulation of these alpha receptors by the pseudoephedrine causes the smooth muscle to contract which, in turn, results in increased obstruction at the level of the bladder neck and results in difficulty emptying the bladder.

Depending on how labile Mr. Stuffy Nose's high blood pressure is, he may also find that his blood pressure is running even higher since he has been taking his Claritin-D for his nasal congestion. The arteriolar walls are well-supplied with alpha receptors that, when stimulated by the pseudoephedrine, cause mild constriction of the arterial walls, and this can, in sensitive individuals, result in an elevation of the blood pressure.

By far the greatest risk to Mr. Stuffy Nose's well-being, as a result of being prescribed a drug having sypathomimetic properties, is the precipitation of a life-threatening arrhythmia or of the decompensation of his previously stable congestive heart failure. The sympathomimetic pseudoephedrine, in the case of the heart, activates primarily the beta receptors, causing the baseline heart rate to increase. In patients with significant symptoms from congestive heart failure or, in the late stages of heart failure, even a small increase in the resting heart rate can result in significant worsening of their heart failure and their symptoms. In fact, maintaining a low-normal heart rate is an important component in the physiological treatment of congestive heart failure. This is part of the reasoning behind a class of drugs known as beta blockers being an important part of drug therapy of congestive heart failure. If you possess a proper medical school education concerning the physiology and pharmacology behind this, it all makes sense, and, if you don't possess such an education,— well, maybe you should not be permitted to write prescriptions for drug therapy. It is in the heart's relaxed phase, between beats, that the ventricles of the heart are refilled with blood in preparation for the next contraction of the heart which will, in the case of the right ventricle, pump the unoxygenated blood to the lungs for oxygenation and the oxygenated blood from the left ventricle out to all tissues of the body in the case of the left ventricle. At higher heart rates the relaxed interval between contractions of the heart is shortened, leaving insufficient time for the ventricles to become entirely refilled with blood for the next beat. This results in loss of stretching of the ventricles' walls and subsequent loss of efficiency of the heart's pumping action. Consequently, this leads to insufficient volumes being ejected from the ventricles during contraction and worsening of the heart's failure.

Thus, neither Mr. Stuffy Nose's case, nor its treatment, turned out to be a simple matter at all. The case only appeared simple to someone having *some* medical knowledge but far short of the knowledge inherent in an MD degree and certainly far short of the knowledge necessary for safely diagnosing and prescribing medications to a patient with this medical history.

Without proper education and knowledge, a license is only legal authority for engaging in an activity. It is certain that, if our health care system is to become the economical, efficient and quality one we are wishing for, it will be necessary to return primary care medicine to its preeminent position in the system. It is just as certain, however, that addressing the now very real primary care MD shortage, via *primary care physician-substitutes* (physician assistants and nurse practitioners), is neither a satisfactory economic solution for cost-control, nor a satisfactory qualitative solution for services being provided to consumers.

Ironically, the entire concept and current irresponsible use of primary care advanced registered nurse practitioners is likely helping perpetuate today's primary care MD shortage. Already the lowest paid of all the medical specialties, their education and training requirements the least understood and therefore least appreciated by public and media alike, the self-images of family practitioners, internists and general pediatricians have taken a beating for the past two decades. Rejuvenation of their self-esteem can no longer even be achieved via attendance of their specialty societies' continuing education courses through fellowship and camaraderie, as more and more frequently, the opening welcoming addresses of such events begin with, " Welcome, *primary care providers,* not with, *"Welcome, doctors"* as used to be the case. Inevitably such a greeting leads to the physicians in the room surveying their fellow attendees and realizing that a significant number in attendance have nursing degrees rather than MD degrees, and the total education requirements and time invested in formal medical science training of these individuals is frequently shockingly minimal and not even in the same ball park as that implicit in the MD degree. Yet, all of them have been haphazardly granted the two privileges, which in the public's mind are most associated with being a doctor, prescriptive authority and the ability to order tests and imaging procedures within the hospital. Under these circumstances, how many primary care physicians today would be likely to recommend to an aspiring son or daughter in medical school that they choose a primary care specialty upon their graduation? How many medical students, after completing the interminable educational pathway to the MD degree and finding themselves one-hundred-fifty-thousand dollars in debt, will be likely to choose primary care as their specialty when a nursing degree, with minimal formal class room education time investment, will confer essentially the same rights and benefits and with pay now rapidly approaching that of a primary care MD as well?

The public needs also to curb its unrealistic expectations of medicine as a science. In a very real sense such inflated misperceptions of medical science's capabilities often serves as a perpetuator of adverse patient behavior or, at the very least, detracts from any motivation the patient might have towards assuming responsibility for maintaining good health. I've had children respond to my admonition that unsafe skateboarding practices could lead to a broken back with the reply,

"Well, I could get it fixed. All I'd have to do is miss school for a few weeks and stay in the hospital awhile." Likewise, there are budding alcoholics out there that, on the occasion they might be reflecting on the need to quit drinking and contemplating premature death due to cirrhosis or cardiomyopathy, enable themselves to continue their drinking by rationalizing,

"Oh well, I can always have a liver transplant or heart transplant."

Likewise, how far should patients' rights go, particularly regarding high-cost and high-risk exotic medical interventions? Does a patient still smoking three packs of cigarettes a day and in need of triple bypass coronary surgery have a right to expect that his insurance will pay for it or that his cardiovascular surgeon will provide it if he intends to continue smoking? And what about the young, obese smoker-patient who has not yet experienced his first heart attack but, having high cholesterol, requests his doctor to change him from his current cost-effective cholesterol-lowering drug to one only minimally more effective but extremely more expensive, because,

"My buddy at work has had a heart attack, and his cardiologist has him on that drug and says it's the best and strongest one. They even say that it's the best one available on TV commercials. Besides, my company insurance will pay for it." Wouldn't it be beneficial and economically responsible if his doctor, without fear of repercussions of any kind, could ask that young man?

"Do you intend to quit smoking? Do you wear your seat belt everyday when you're in your car? Do you exercise at least three days a week?" Are you compliant with your low-cholesterol diet?" Do you intend to pay for the medication out of pocket, or are you going to turn it into your insurance company?" Should a physician have to be worried in such circumstances as these that, if he refuses the patient's request on the grounds that he is still smoking and noncompliant with his diet, the patient will simply find another doctor who will gladly give him what he wants if he might get a new patient out of it. This is the practice environment today, thanks to direct advertising to the public by the pharmaceutical companies and to increased competition for the patients with the ability to pay. Physicians today are so hamstrung by frivolous lawsuits, patient rights advocates and fear of losing dissatisfied patients to their competition that they can no longer demand personal responsibility as a health maintenance tool for their patients. All parties would benefit if physicians were able to still require of their patients, personal responsibility in their own health maintenance.

Today's healthcare consumer has need to know that, for every one Mercedes driven by a physician, there are four more on the road paid for also with consumer healthcare dollars but having non physician types from medical management business corporations behind their steering wheels.

In the sixties and seventies, the cost of medical care to the patient was pretty much represented by physician fees and hospital charges. The overhead in those days was simple and similar to the overhead of any other business. Today, the cost to the consumer is much more illusive to accounting due to a legion of hidden costs within the now exorbitant administrative overheads of physicians' offices and hospitals. With the single exception of increased expensive technological advances, none of these hidden overhead costs have anything to do with actual healthcare delivery by a physician.

Three of the monster contributors to the high cost of medical care today are physicians practicing defensive medicine, maintenance of the paper trail made necessary by Joint Commission and an abundance of healthcare providers (increasingly, non physicians today) for inappropriately ordering increasingly expensive diagnostic tests. Simply put, there are more people able to charge for and be reimbursed for healthcare delivery than ever before, and, increasingly, their charges are far out of proportion for their formal educations and their worth to the system. This has come about through carving out of niches for specific services. As each niche is found to be profitable, others jump on the same band wagon, and, as the qualifications necessary for being part of a particular niche become defined and in demand, suppliers of the *minimal* qualifications required crank up for meeting the demand. Frequently, in their haste to do so, they lower their admission standards, educational requirements and the total educational time investment. This certainly appears, from my personal observations and experiences, to have been the course which the 2-yr. RN programs have taken, and now many of the advanced nurse practitioner programs appear to be following the same negative educational evolution, as their popularity, salaries, and numbers continue growing. The end result of previous such cycles has been an oversupply of niches occupied by persons commanding incomes far out of proportion to their actual education, skills and overall worth to the health care system. The health care persons staffing and being remunerated in these niches are, for the most part, RNs and advanced nurse practitioners. For the patient, whatever the niche being utilized, this frequently translates quite simply into being seen and managed by someone having much less education than an MD but at a fee approaching that of a physician's. Of even greater importance, with the baby boomer health care needs on the horizon and Medicare coffers dwindling, is the addition of so many of these cost-inefficient hands into the Medicare cookie jar.

Presently boutique niches are beginning to take off within hospitals and health management businesses. With competition for markets at an all time high and, already into the home healthcare delivery aspect, hospitals and healthcare businesses have expanded their niches even further with "incontinence centers," "wound healing centers," "sleep apnea centers," "pain centers," "colostomy care centers", and others are certain to follow. One of the most incredible niches of them all has been the success that Hospice has had in making care of the dying a specialty niche!! Has Medicare bothered to check recently the amount of money they are paying out to Hospice for assurance that *dying* is being managed correctly?! The work-horses being remunerated for staffing most of these niches are RNs and ARNPs, with a physician usually somewhere in the distant background getting a cut of the action for his/her alleged supervisory service. Each of these niches must have support staff and other overhead, all of which their charges must be sufficient for recovering. In the fifties, one doctor provided all the services that the niches provide today. In the sixties and seventies, one appropriately motivated primary care doctor, using the occasional consulting specialist, provided the same services. Today, we have entire teams representing each niche. In the fifties, a single physician's livelihood was earned providing these services, and today, three to six incomes are represented within every boutique niche. Multiply six incomes by the number of niches already being foisted on the public as necessary for *boutique* care, and we are talking about significant bucks indeed! Since these niche centers frequently are located within and supported by corporate managed hospitals, they benefit from hospital referrals and hospital marketing to the public. Sadly, it is becoming increasingly common today that patients, not referred by their personal physicians for the services provided by these niches, mistakenly assume they have not had good care. This is not so, as the care from their personal physician frequently is better and more economical as well. In actual fact, many of these boutique niches are often developed by hospitals or physicians for the express purpose of supplementing their revenues.

If you remain still with doubts concerning my economic observations, then take the time to stroll through the parking garages of Humana, HCA and all the other healthcare management corporation offices. While doing so, count the number of Volvos, Mercedes, BMW's and other expensive rides. Also, crane your neck upward toward the distant sky and enjoy the high-rise architecture of the affluent. Remember, while on your visit, that these are only the largest grandfathers of the industry, and there are legions of their children and great-grandchildren dotted throughout our nation having smaller but no less successful and elegant offices. Inability of locally appointed hospital boards to keep up with increasingly complex bureaucratic edicts has brought us today to a multibillion dollar industry where Volvos, Audi's and Mercedes abound. Every dollar these folks make comes through managing a physician or hospital and is passed along by physicians and hospitals to the patients or patients' insurance carriers. In the sixties, we had only physicians and hospitals submitting charges to patients. Today, we have an entire multibillion dollar health related business sector of our national economy surreptitiously also charging the patients. In the sixties, we had a medical profession consisting of professionally motivated physicians. At the millennium we have a medical business where most physicians are merely highly educated employees, motivated as most employees, by survival, working conditions and the best remuneration possible.

The recent health care reform debate has certainly been an emotional roller coaster ride for me. More than one occasion has found me shouting out at the TV, "None of you really gets it!!" Yes, the cost has to be gotten under control if the country's economy is going to survive, and yes, there are very real issues regarding millions of folks unable to afford access to health care, but what about the issue of quality? There are some very real quality problems here as well, and they seem to be flying entirely under the debate's radar despite the fact that, to a very large measure, *poor quality of care is the cause* of much of the system's run-away costs. On my most cynical days, whenever I hear a politician or pundit ranting,

"We must get the thirty-million uninsured access to our health care system," a part of me wants to yell back at them,

"By all means let's get them covered and into the current system too, so that they can contribute their fair share to the 100, 000 or so deaths per year in the nation's hospitals due to medical errors and other quality of care issues."

Any time a patient in a doctor's office or a hospital receives a test or a treatment that really wasn't necessary, the quality of care is bad and the cost of care goes up. Today, it is likely that at least a fourth of all patients seeing a health care provider for an acute problem will receive a treatment or testing which was either not needed or not effective. A patient going to their personal physician's office and being diagnosed with an uncomplicated case of the flu, leaving with a prescription in hand for an antibiotic, has experienced poor quality of care and two needless expenses, since influenza is caused by a virus and antibiotics are useless against viruses. Yet, inappropriate prescribing of antibiotics during the flu season is a common occurrence throughout our health care system.

A patient visiting a physician to complain of classic signs, symptoms and history of a migraine headache, whose neurological exam is entirely normal, and then leaves the doctor's office with an appointment for an MRI of the head, has experienced bad quality of care and significant needless expense. A competent physician should not need the expensive imaging to diagnose a migraine headache under these circumstances, and the doctor ordering it did so for self-serving reasons having nothing to do with better case management.

A patient in a hospital for an elective surgery who leaves the hospital with a urinary tract infection, caused by improper nursing technique during insertion of a Foley catheter, did not receive quality care. Likewise, a patient in the hospital, given an antibiotic before surgery prophylactically (to protect against possible infection expected with that particular type surgery), who subsequently winds up with C. difficile colitis (a serious infection of the colon occasionally occurring with powerful antibiotic usage), when the surgery really didn't meet evidence based standards for necessitating antibiotic prophylaxis in the first place, has experienced poor quality of care.

The patient with a severe infection of the colon from inappropriate use of an antibiotic and the patient going home from the hospital with a urinary tract infection she did not have when she entered the hospital, both experienced adverse effects to their physical well being, and, even should they not immediately make the connection to poor quality care, they easily could be made to understand the connection with a proper explanation. Eventually these two will likely make the connection and realize that they received poor quality of care. However, the patient given a useless and unnecessary antibiotic for the flu and the headache patient undergoing an expensive and unnecessary MRI of the head for her doctor's comfort and convenience, did not sufferer any physical harm as a result of the poor quality of care they received, and the two of them likely will never even realize that they were recipients of poor quality of care. Indeed, the patient given an antibiotic of no possible use against the influenza virus will eventually recover from the flu anyway, probably even erroneously crediting the useless antibiotic with her recovery, rather than to the self-limited nature of the flu as was actually the case. Likewise, the headache patient, paying dearly for an unnecessary MRI of the head because of her physician practicing defensively, probably was treated by the Dr. for the Migraine headache he already knew she had before ordering the MRI, and, now grateful that she is better, this patient might even mistakenly extol her physician's "thoroughness" for having ordered an MRI of the head. The point to be taken is that poor quality of care and excessive costs are intimately related, and both are much more prevalent in the current system than anyone imagines and go entirely unrecognized by consumers, except in instances which result in significantly bad medical outcomes, which although disturbingly numerous, are still outnumbered by instances in which poor quality of care results only in needless and excessive expense for the patient. Make no mistake about it. There is plenty of waste and excess cost due to poor quality of care and self-serving decision making by clinicians within our health care system. If this could be rooted out and kept out by a meaningful monitoring and punishment/reward system designed for changing the guilty doctors' behavior, the economics of our health care delivery system would change dramatically for the better.

The current health care delivery debate seems blind to the quality problems within the current system, and, paradoxically, whenever the issue of quality has been raised during the debate, it usually is as a point of pride and framed in the favorite and recurrent battle cry of some,

"We have the best health care in the world. Why should we want to change it?" In reality, what we have is an abundance of the best medical expertise and an abundance of the best and most expensive medical technology in the world. The problem is that, increasingly, we also have an abundance of medical expertise who, under the influence of business pressures and cultural expectations, is no longer ordering tests and treatments in the manner they were trained en route to their prestigious diplomas and specialty certifications. This has resulted in routine over-utilization of some of the most expensive diagnostic technology in the world and, all too often, employment of medical interventions and treatments that are of minimal to no benefit to the patient— all because of their availability and the consumers' desire for and expectation of them.

Currently the consumer contribution to costly and poor quality medical care, via their unquestioning acceptance of ineffective and futile treatments, equals or surpasses that attributable to the self-serving clinical providers. This is occurring due to the public's medical science misperceptions and their unrealistic expectations. The best solution for this is a practical medical education for the consumer. This will necessarily be a lengthy and difficult process, but it can no longer be put off, and the sooner the process is begun the sooner the eventual benefits to the health care system's cost-control can be accrued. Making the public more savvy in matters of medical decision making will be extremely difficult, as any reputable physician following the recent health care debate exchanges among purportedly very educated, but non physicians, from both sides of the issue can attest.

For example, I recall on one occasion listening to two erudite and highly respected political pundits attempting to express their concerns on the potential horrors likely to be associated with any rationing of health care for the elderly. Each implied a worst case scenario in generalities, but neither gave specific clinical information sufficient for even the most astute and experienced physician to arrive at a best, patient-interest decision with. Perhaps these news personalities did not have more specifics on the cases they referenced, perhaps they had more specifics on the cases but withheld them for the purpose of *spinning* their cases to make their points, or just maybe, like most of the general public, they did not know what they were talking about medically.

One gentleman, in attempting to make his point, referenced a family member in her nineties who had undergone a "cardiac procedure" that, had it been denied her due to her age, might have cost her life in his opinion. With no more specific clinical information than this it is impossible to say whether this case exemplified excellent quality care or bad quality care. To be sure, part of the answer depends upon whose point of view you are assessing the care from. The person's view point of paramount importance is the patient herself. If this patient, at ninety-plus years, was still cognizant, enjoying interaction with her environment and wished the procedure done so that she might continue to enjoy her life further, then humanity and common sense dictate that having the procedure done represented quality health care. If, however, the patient was a ninety-year-old, total-nursing-care patient with no capacity for recognizing or interacting with her environment and no chance of changing that fact, regardless of the type of medical intervention employed, then the decision to have her undergo a cardiac procedure for the sole purpose of prolonging *physiological life* would not have been in her best interest. Performing a major medical procedure under such circumstances would represent poor quality of care in the minds of most medically objective and compassionate people. Indeed, if it were possible for such a patient to suddenly return to her full mental capacity of five years previously for witnessing her present condition and the offer being made for a cardiac procedure for the purpose of prolonging only *physiological* life, the odds are overwhelming that she would want no part of it. The author's experience has been that the majority of cognitive, geriatric patients, when presented with similar clinical hypothetical end-of-life scenarios, opt for no further medical intervention. Geriatric studies in the medical literature have corroborated this. One study even found that a surprising number of successfully resuscitated geriatric patients, upon being interviewed much later regarding the experience, said they would not want it done again; remarkably, this group still had cognitive function and were not nearly as lacking in life-quality as the hypothetical patient presented here! The pundit, in referencing his ninety-year-old family member's cardiac procedure, provided insufficient information regarding the family member's overall health status for the audience to decide whether the cardiac procedure was a good thing for her or a nightmare. This reference to his ninety-year-old family member's having access to a cardiac procedure would have been of much more educational value to the viewing audience if the issue of meaningful life vs. *physiological* life could have been explored further, since cardiac and other major medical procedures are being done all too often today on patients already possessing only *physiological life,* and, sometimes, such procedures are being performed within days to weeks of the patients' inevitable and expected natural deaths. This practice represents a needless and unnecessary additional burden and source of suffering for patients in these circumstances. It also represents a needless and extremely significant financial burden on private-pay patients and, in particular, on Medicare's dwindling coffers.

Each end-of-life case is different, but when the elderly still in possession of their cognitive skills are asked their views on medical intervention indicated for their peers occupying the worst-half of the illness/quality of life spectrum, there is usually general consensus among them favoring no further intervention except for comfort and dignity measures. Unfortunately for end-of-life geriatric patients, their personal wishes often get cast aside in the last weeks to months of life when decisions end up being made by doctors, nursing staff, nursing home administrations and families— all parties making their decisions based on self-serving agendas, usually convenience or defensive medicine for the clinical group and guilt or misdirected grief in the case of the families. Indeed, a good argument could be made that geriatric patients often would be better served by randomly selecting any objective, compassionate adult of common-sense for making their end-of-life care choices for them. This personal view comes after thirty-five years of dealing with family guardians, legal guardians, state guardians and court appointed guardians, with the last two categories generally tending to be the most self-serving of all, in my experience. A blessed exception is the small family or family member guardian who has maintained continuous periodic contact for observing their loved one's transition from meaningful life to physiological life with all its accompanying indignities and suffering. These folks simply follow *The Golden Rule* and serve their loved one's best interest extraordinarily well. In many medical matters in general and particularly in end-of-life care, doing nothing often represents quality care, while doing "something" frequently leads to poor quality care, unnecessary suffering and an increased financial burden to the cost of health care.

Also during the recent health care debate news coverage, I recall another respected political pundit referencing a cancer patient in Canada allegedly having had a life-prolonging cancer therapy drug denied them under the Canadian system, the implication being that it was denied due to the drug's high price tag. Certainly drugs within this therapeutic category generally are very costly indeed, especially the newest and most recently developed of them, and this points out the importance of knowing all the facts and begs further questions as well. For putting this reference into any meaningful perspective regarding the health care debate, one would at least need to know the name of the drug in question and, most important of all, a precise and plain-speak description of the drug's expected benefit to a patient of identical age, cancer type and overall clinical status.

What, in the interest of drug-consumer education, if the denied drug by the Canadian Health Care System in this reference was the USA's recently FDA approved *Avastin* for the treatment of breast cancer? If it were that drug the pundit was referring to, there are many medical researchers and physicians in the U.S. who should applaud the Canadian Health Care System's decision or risk being hypocrites in the extreme, since the FDA's approval of the drug in the U.S. surprised many in cancer research and within the medical field. The FDA's decision in the case of this drug was a controversial one and represented a change from their previous standards followed in approving drugs of this class. It would appear that data presented to the FDA by the drug's manufacturer did not document either prolonged survival or improved quality of life with use of the drug. Prior to this instance, the FDA had required one or the other of these criteria be met for approving a new agent within this class of drugs. In fact, the FDA's own physician-advisory panel voted not to approve the drug. The FDA, however, surprisingly chose in this instance to administratively override the decision of its own advisory panel of physicians, granting approval anyway. The decision by the FDA to permit a new and extremely pricey drug to be sold to consumers, without any hard scientific evidence that the drug would be of any meaningful benefit to patients, simply fuels an increasing chorus of voices questioning the quality of work and/or motivations of the FDA. Indeed, it would be well-worth the general public's time to familiarize themselves with the entire process required for FDA approval of a new drug for market, particularly scrutinizing their standards for proof of efficacy i.e., how well the drug really works, not couched in the scientific speak of "statistically significant", but simply and practically what the drug can be expected to do for the majority of patients taking it. The standards required are different depending on the class of drug and whether the drug represents a first agent in an entirely new class of drugs or is only a new drug within a class from which a prior drug or drugs have already received approval. My personal experience in attempting to find specific and definitive FDA efficacy standards required of new drugs through an on-line search has been unrewarding. While specific standards may exist, I had no success in locating them. Much of what I did encounter in the search, however, leads me to strongly suspect that that the baseline against which new drugs are being measured may be *placebo,* and the standard may well be that the new drug need only be proven to perform *statistically significantly* better than placebo i.e., better than a "sugar pill". If this indeed should be the case, how the words *statistically significant* translate practically into observable clinical benefit to the patient, of course, then becomes the sixty-four-thousand-dollar question for the consumer or insurance company paying for the drug. With little to no emphasis being placed on *how much better* a new drug works than a "sugar pill" or better than previously approved drugs of the same class with which it is being compared, a potential situation is created for consumers where, for example, newer and pricier blood pressure medications, having only minimal effect on lowering blood pressure, could possibly be aggressively being advertised by their manufacturers to physicians and consumers, when some of the older, less advertised and less expensive blood pressure medications, already available, are considerably more effective.

It takes little brain power for figuring out who benefits most from this standard, especially given the fact that it likely costs the drug manufacturer much less to develop and obtain FDA approval for a new drug in a class already having prior approved drugs within it, than it costs for researching and obtaining approval for a new drug which is the first agent in an entirely unknown and pristine class. This is an economically very important bit of information for patients given the multimillions of dollars being spent by drug companies annually on direct consumer advertising of their *newest* drugs and the fact that their *newest* never cost the same or less than their same-class older prototypic drugs. Occasionally, the *newest* in a long line of previously approved drugs of the same class might offer some subtle clinical advantage over older drugs in its class for carefully selected individual patients, but only a well-educated, properly motivated physician of integrity can provide that information for the consumer. Very frequently, however, older prototypic drugs of the same class that already are on the market will perform equally well or better than the newest agents out. Indeed, an argument could be made that some of the older drugs are safer, given their prior years of use for the collection of safety data.

If Congress were really interested in lowering health care costs and looking out for patients' best interests, they could begin with a thorough revisit of the laws that they provided the FDA concerning approval of new drugs for market. Simply raising the bar on efficacy standards for all new drugs would save millions in health care costs. After all, if you had been using brand "XXX" toilet bowl cleaner for years and paying three-dollars for a 20 oz. container for it, is it likely you would try a new brand of toilet bowel cleaner showing up on the retailer's shelf that sold for <u>seven</u> dollars per 20 oz. container, if its sales pitch was that it "works just as well as "XXX" toilet bowl cleaner"?!

For the sake of making an educational point for medical consumers, let us assume that it was *Avastin* being referenced by the political pundit during the health care debate as having been denied the patient with breast cancer by the Canadian Health Care System. At nearly $8,000 cost per month and with the data from its one clinical trial showing no improvement in quality of life or in survival, surely no person of common-sense could think that permitting a drug of such unproven benefit to be marketed to physicians and consumers, with the expectation that medical insurance companies, Medicare and Medicaid should be willing to pay for it, makes any economic sense whatsoever. Yet it happened! This is a prime example of where government, leaning towards health care consumers rather than toward unbridled capitalism and special interests, could have been mutually beneficial for patient care *and* health care cost control. Indeed, patients would have been no worse off, and insurance carriers and Medicare would have been much better off, had the drug's approval been left up to the Better Business Bureau in this particular instance.

Attempting to assign an acceptable dollar-amount for any quantifiable additional time among the living is a nightmarish fool's task regardless of how positively glowing a treatment's quality and survival data appear, but, in light of the clinical trial data in the case of *Avastin,* it would seem like a "no brainer"; certainly it would be so for all but the most irrational and extravagantly wealthy of consumers. Whether the drug alluded to by the pundit during his opinion piece as having been denied the patient with cancer by the Canadian Healthcare System was *Avastin* is really immaterial, except, perhaps, for serving to remind us just how naïve and vulnerable consumers are to the direct marketing of the latest drugs. The point plainly is that there were thousands of consumers who, after viewing the TV coverage and listening to the pundit's general accusation of a drug that, in his opinion, was extremely beneficial but being denied a patient in the Canadian Health System due to its cost, would likely be appalled and outraged that a government could be so callous. As a physician, aware of the recent approval of *Avastatin* for the USA market and the concern and controversy it generated within the medical community, my take on the pundit's story, however, was much more cautious, if not downright suspicious. While it is entirely possible that the pundit was referring to some drug other than *Avastin,* the point is that there is much the public doesn't know regarding the real and practical benefits to be expected from many of the drugs being vigorously marked to them, and the non physician public is ill equipped for sorting it all out and remain woefully vulnerable to advertising and media coverage of new drugs and the latest medical "breakthroughs".

Make no mistake about it. There is more than enough inefficiency, sloppy medical-decision making, profiteering and overuse of expensive technology within the current health care system that, if immediately eliminated, would result in the quality system we all wish for and in savings sufficient for welcoming those without coverage into an improved and better quality system, putting Medicare back on the road to solvency in the process. This can be achieved but only if the government finds the heart and political will for standing up to special interests and for intervening in specific instances where unbridled capitalism is contributing in equal measure to poor quality of care and to out-of-control costs within our health care system.

For weeding out the expensive and poor-quality practice habits increasingly pervasive among physicians in the system today, the government would do well to remember that "it takes one to know one,", and, in this case, it will certainly take *some* to correct *some.* The government will need as its committed ally for this formidable task, a sizeable and committed representation from the nation's practicing physicians. While the numbers of this group should be sufficient for representing the nation geographically, total numbers are not as important for the success of its mission as would be the group's prevailing medical philosophy including: commitment to evidence based medical science, integrity and common-sense. In short, government will need as its advisory ally only the most scientifically and medically righteous physicians for assuring the mission's success. Physicians qualifying for this committee should share many of the same characteristics and principals which defined their predecessors during the period when there was still a noble *medical profession.* These should be individuals who went to medical school because of their passion for the medical sciences and for the sheer fascination with applying them practically for diagnosing and treating illness. These should also be the folks who take the Hippocratic Oath seriously. The vetting process must be meticulous, lest the committee should become contaminated from the ranks of their physician counterparts having more self-serving medical philosophies and ambitions, thus rendering the committee impotent. Hastily comprising a roster of the usual names from medical academia will not be sufficient for keeping out special interests or for achieving much-needed, real-world practicality and effectiveness. Sufficient time should be taken for truly getting to know the hearts, ambitions and medical philosophies of every individual before their appointment, since this will be the committee who will help the government identify and bring to task the costly, prolific and self-serving decision makers among practicing clinicians in the current health care system. This should also be the committee who helps the government develop clinically sound and cost-effective formularies, clinically effective and cost-effective standards of care and helps define and educate the general public as to what constitutes *futile medical intervention* in the context of a number of commonly encountered clinical situations that are, due to self-serving clinical decision making, being routinely mismanaged throughout the health care system currently. Once meticulously vetted, this working committee of physicians must be empowered by Congress with developing medically and ethically sound standards for all of America's practicing physicians. In short, this committee must strive to fill the void left after cultural and business interests led to the premature death of the proud and noble *medical profession!*

In their attempt at bringing much-needed cost control to the American health care system, Congress faces the unique conundrum that the perpetrators of much of this unnecessary cost are the American consumer and the American ideology, *free enterprise.* The unpalatable truth is, for any real and long-term cost control to be achieved, Congress will need to make a calculated special exemption in the case of health care and veer slightly away from total sacred-cow status for free enterprise. Initially this would be met with angry rhetoric and doomsday predictions from the drug manufacturers and the medical sector's lobbyists, but it might not be as bad as Congress fears. Both the pharmaceutical industry and the general medical sector are comprised of savvy and practical people, and such individuals should easily understand, without meaningful cost control in the current health care system, their business worlds are eventually likely to crash and burn anyway, at the hands of draconian income tax hikes or at the hands of a total single-payer government system hastily and emergently passed for addressing a national economic crisis precipitated by our inability to compete in the global market due to the continually rising cost of health care.

By proactively dampening, only slightly, the rampant free enterprise philosophy within the health care sector, tremendous dividends in cost reduction and improved quality of care could be achieved simultaneously, since being sold tests, drugs and treatments which are unnecessary or of marginal effectiveness, constitutes poor quality of care. Towards these ends Congress might begin with the following measures: provide a meaningful and cost effective replacement for the Joint Commission; an immediate ban on all direct advertising to consumers by pharmaceutical companies, hospitals/clinics, and physicians; meaningful torte reform sufficient for immediately eliminating the perceived need to practice *defensive medicine;* permission for health insurance companies to engage in interstate competition; resurrection of a real and meaningful certificate-of-need process for any additional new hospitals, surgical centers, dialysis centers, physical therapy centers, imaging centers and for all expansions within pre-existing facilities among these same categories.

Finally, a rewrite of the laws governing the FDA's approval process for new drugs is needed. The new laws should emphasize raising the bar on efficacy standards for new drugs. Doing so would afford meaningful consumer protection and bring down the cost of care as well.

CHAPTER 6

THE AUTOPSY

All aforementioned suggestions would make today's medical business more efficient, more effective and more economical, but it won't bring back the medical profession of the past. In the sixties, Medicare entered the arena and physicians, although extremely wary of potential government involvement in the long run, still saw it as a way of recouping dollars that were being carried on their books for services rendered to the elderly who could not afford to pay. Hindsight reveals the eventual acceptance of Medicare by the medical community to have been a fatal blow to the future quality of our health care system and one which birthed a self-perpetuating, paper-trail-monitoring system that today has evolved into an entire multibillion-dollar business sector hidden within the cost of care for today's healthcare delivery. It includes thousands of consultants, accountants, attorneys, hospital management firms' corporative staffs, nurses, physicians, pharmacists, dieticians, and risk managers, to make just a partial listing. Many of these are full-time, on-site hospital employees, while others are privately contracted services to the medical facilities. They are legion in numbers and diverse in their services, but the common denominator for them all is that their jobs are predicated on helping hospitals and clinics meet the complex bureaucratic demands of the Joint Commission and other regulatory agencies or, more accurately, helping hospitals generate the perfect paper trail for *appearing* that the complex demands of these agencies have been met, so that the almighty-Medicare dollar flow will not be interrupted. The paychecks of these thousands of individuals, whether issued by the hospital or issued in the private sector, represent millions and millions of dollars which is passed right on to the patients through higher charges for services and higher insurance premiums. In 1993 the *New England Journal of Medicine* published a 1990 nationwide survey of hospital administration costs which showed administrative costs accounting for

24.8 percent of hospital expenses, more than twice the amount for Canadian hospitals. What must current figures look like!

Shortly after Medicare's debut, the high-technology era of medicine was ushered in by the CT scan in the seventies and, following quickly on its heels, came fiber optics, laser, and a legion of other terribly expensive techniques and devices. Suddenly, doctors' overheads shot upwards, and they, in turn, countered with sudden increases in their fees. As patients began to smart from these first jumps in a heretofore stable price arena, they began to clamor about it and to ponder it as well. For the first time some of them took note that their physicians drove expensive automobiles and lived in large homes. Many became resentful with what they saw as high-lifestyles, all the while overlooking the fact that the demanding work hours of physicians, prior to the eighties, kept most of those docs from ever being able to enjoy those nice cars and large homes. Aware of the public's concern and determined that no federal dollars were paying for nice automobiles and large homes for physicians who were not doing their jobs or to hospitals providing poor service, Medicare's knee jerk reaction was to increase their paper trail monitoring requirements and, in the process, force the nation's hospitals to dedicate the hourly wages of more hospital personnel to the maintenance of the paper trail. The individuals necessary for dealing with this paper trail task were borrowed from the most degreed and experienced nursing staff within our nation's hospitals, and, as a result, the quality of the clinical care being provided the patients suffered a negative downturn that still haunts us today. As with any business dependent on federal dollars, there may have been some abuse of the Medicare system by a few physicians and hospitals. However, the amount involved in those limited instances pales by comparison to the dollar figure added to the total cost of healthcare delivery by the ineffectual and meaningless paper-trail overkill being carried out today for CMS (the federal agency administering Medicare and Medicaid services) at the hands of their surrogate, the Joint Commission.

This period of time provided opportunistic business types the ideal melieu for taking advantage. So-called medical management corporations rose to the occasion and convinced the hospitals that they could provide the expert resources for understanding and successfully dealing with this paper trail nightmare while increasing hospitals' bottom lines at the same time. Two decades later the same hospital management giants, now fat from hospital management revenue, went, for a while, after physicians' practices, convincing doctors that they could free them from worries of office and clinic administrative tasks, leaving them to simply practice medicine as they loved. So, by the eighties, the medical management business world was well entrenched in our healthcare delivery system. Those in the business of medical management of our nation's hospitals and clinics have been prolific and profitable. Today they represent a very significant portion of the nation's economy and likely are responsible for billions of dollars of additional cost of care within the nation's health care system annually. Politicians and the public, already jaded toward doctors from their continuous bombardment by television dramas depicting the over-dramatized lifestyles of physicians, were pushovers for the sales and publicity departments of these seasoned business moguls, and they quickly bought into their nonsensical pitches that they could halt the out-of-control healthcare costs without any resultant sacrifices in quality or services. Anyone still believing these promises needs only walk the sidewalk in front of the towering, monumental, high-rise of the Humana Corporation Headquarters, now the pride and joy of Louisville, Kentucky's skyline or in front of any of the other corporate headquarters of medical management firms in the U.S. today and simply think about it.

In the early seventies, I was walking between rows of cars in the doctors' parking lot at the hospital when I chanced to overhear a man commenting to his wife as they walked past the lot,

"Look at that would you? Every one of those doctors drives a fancy, new, big car," he disdainfully spat out. It wasn't something I hadn't had presented to me before, although usually more in teasing fashion. However the manner though, it always hurt my feelings and angered me. If any of us had possessed an inkling of how demanding and difficult our personal lives were going to be as physicians, we might have given more thought to clinical research over private practice. Sleep-deprived, never at home and late or absent from every family event were the norm for all physicians back then. The time we got to spend in those "fancy big new cars" was negligible and consisted mostly of driving to and from the hospital. Two decades later while attending a medical meeting in Louisville, Kentucky, as per my passion, I read a dining review on a very upscale, continental restaurant. The menu, décor, ambience and chef's credentials were given glowing accolades by the newspaper's restaurant critic. I looked it up by street address and, finding it to be only six blocks from my downtown hotel, decided to walk to it even though it was already dark. Glad I'd worn my best suit, I thought I had bitten off more luxury than desired when I arrived to find a doorman in tails, Maitre d', wine steward, soup and salad chef, entrée chef, dessert chef, elegant surroundings and a perfect culinary experience. Imagine my shock, when perusing the restaurant's history provided on the back of its dessert menu, I learned that the restaurant was conceived and developed originally as the private dining room for the Humana Corporation executives (one of the nation's largest hospital management corporations), and it only later opened to the general public!! Dark or not, upon leaving that elegant restaurant, I turned sharply to the left and craned my head upward into the night's sky to view a gleaming, multi-storied, stone, architectural, jewel of Louisville, Kentucky's skyline— The Humana Corporation Building! The next day I walked back to that entrepreneurial megalith to be certain that its accent lighting the night before hadn't rendered it greater in the dark than in the light of day. It hadn't! Eventually I made my way around its base to its associated parking structure, and do you know what I saw within it— lots and lots of really "fancy, big, new cars"! None of these cars, however, were paid for by performing a life-saving operation, delivering a baby, treating pneumonia or relieving a patient's pain. Indeed, none of those extravagantly expensive rides were driven there by people who had even so much as placed a band-aid on a patient's cut finger, wrapped a sprained ankle or provided comforting verbal reassurance to a patient. The people who had driven all those fancy, big, new cars to work and now sat, calculating away inside that architectural masterpiece, did not practice medicine nor do any work for patients. Instead, it was the patients, who were working for them— thru paying the inflated medical care costs as a consequence of the handsome salaries being pocketed by the multitude of administrative types employed within that architectural monument to business machinations. Standing there and staring at the premium vehicles housed in that building's garage, my mind slipped back years ago to the man's snide remark regarding doctors and their fancy new cars. If I'd had a clue as to the man's name or how to locate him, I would have called him right there on-the-spot and read him a list of the Mercedes, BMW's, Acura's and other high-priced rides parked within that garage, reminding him that the gleaming, gargantuan building which that parking garage served was not a medical clinic, not a hospital and did not contain a single practicing physician within it. It was, however, built with consumer health-care-dollars, nonetheless.

Medicine, dead as a profession, is alive and well today as medicine *the business*. With each passing year the ranks of the physicians who trained and practiced when it was a profession are increasingly being diluted by a generation of medical school graduates whose motivations, ambitions and goals have been, influenced by and developed within, the era of *medicine the business*. Never having known the pride and ethics inherent when we were a profession, these new-age docs cannot be expected to be satisfied with measuring their self-worth and practice successes via professional esteem bestowed upon them by professional peers. Forced by economic and cultural factors into following a business model, younger MDs increasingly act as businessmen and women, looking to working conditions and remuneration as their primary means for measuring their personal success. After all, if you look like a business and are treated like a business, sooner or later you're bound to behave like a business.

The medical school diploma that past generations of physicians obsessed over, lusted for, prayed for and coveted as the holy grail of the most exclusive fraternity/sorority in the world, has become just another necessary license for engaging in a specific commerce in order to maintain a now sought-after lifestyle. *Caveat emptor* should be the patient's credo from here on out. Today, patients would be well advised to remember that degrees, titles and diplomas, in the absence of meaningful education, integrity and professionalism, are merely licenses for doing business, no matter how impressive the documents themselves might appear. From here on out, the consumer should neither be surprised at where scientific, quality-medicine occasionally will still be encountered, nor at the *big-name places* where it will be conspicuously absent.

On the week that the first draft of this writing was completed, June 21, 2000, the World Health Organization ranked the world's health care systems by country from best to worst. They ranked the United States 37th down the list, and the media today continues to be filled with stories of spiraling hospital deaths due to medical errors, spiraling health care cost and spiraling numbers of those who cannot afford health insurance for even accessing the wasteful and increasingly quality-plagued health care system we currently have.

Author Biography

The author was born in Du Quoin, Illinois July 5, 1947. As a young boy growing up in Du Quoin much of his time was spent in a drugstore where his father was the only pharmacist. Two general practice physicians occupying the second story of the building often took their breaks at the drugstore's soda fountain, affording a young boy ample contact time for developing a respect and admiration of physicians which early on charted his course toward an M. D. degree.

Dr. Cato earned a B. S. degree with majors in biology and chemistry from Western Kentucky University in 1969. While at Western he was inducted as a charter member into Alpha Epsilon Delta Honorary Premedical Society.

Dr Cato received his M. D. degree in 1973 from University of Louisville School of Medicine and in 1974 completed a rotating internship at St Mary's Hospital in Evansville, Indiana. He is a diplomat of the National Board of Medical Examiners and a diplomat of the American Board of Family Practice. Dr. Cato is also a fellow of the American Academy of Family Physicians.

The author's 34 years of practice as a family physician chronologically spanned the two decades in which his proud and noble profession was transformed by business interests and cultural forces into an inefficient and bottom line orientated medical business.

During his 34 years of practice, Dr. Cato engaged in private family practice in small towns in two different venues. The first of these was a southern Illinois community of 7,000 people and the second was a Kentucky community of 4,500. The two communities were of entirely dissimilar demographics and experiences, which combined to provide the author with a broad and unique perspective regarding the problems with medical care as it currently is being delivered today.

In 1988 Dr. Cato left the private sector and became the hospital-based Medical Officer for one of Kentucky's 222 bed state psychiatric hospitals where his responsibility consisted of the management of all non psychiatric acute and chronic medical conditions of the patient population within the hospital. Additionally, he served as Medical Director and full time attending medical physician for the hospital's on-grounds 120-bed skilled care/psychiatric nursing facility. Dr. Cato retired from this position in November, 2008 after 20 years of service.

Along this medical journey Dr. Cato also served:

As physician member of county health department boards in both Illinois and Kentucky

As physician in charge of the county health department's Child Immunization Clinic while in Illinois

As company mine physician for Amax Coal Company's under ground mine while in Illinois

As physician member of Wabash General Hospital's board of directors in Illinois, which, at that time was under the management of Hospital Corporation of America (HCA) in HCA's nascent years as a pioneer in the nation's burgeoning and, soon to be, lucrative hospital management industry.

www.ingramcontent.com/pod-product-compliance
Lightning Source LLC
Chambersburg PA
CBHW030922180526
45163CB00002B/431